Agentic AI

A Wiley Brand

Agentic AI

by Pam Baker

Agentic AI For Dummies®

Published by: **John Wiley & Sons, Inc.,** 111 River Street, Hoboken, NJ 07030-5774, www.wiley.com

Contents at a Glance

Table of Contents

PART 4: EXPLORING MYTHS AND REALITIES

CHAPTER 11: Dispelling Common Agentic AI Misconceptions

CHAPTER 12: Upskilling for the Agentic Age

Introduction

As of early 2025, industry consensus estimates that 115 to 180 million individuals worldwide have been using artificial intelligence (AI) on a daily basis, and you probably have encountered it yourself. Maybe you asked a chatbot to write an e-mail for you, or you used an AI image generator to make a funny picture of your dog in a spacesuit or to make changes to a photo you took with your phone. That kind of AI — which generates text, images, or sound from nothing more than your spoken or typed command — is called *Generative AI* (GenAI), and it's been all the rage since late 2022.

But AI is constantly evolving, and the new wave is called Agentic AI. The term *agentic* simply denotes AI that doesn't just sit there waiting for you to enter a command in the prompt bar. Instead, it can take action on its own accord. It follows goals and the framework that you set, but it finds its own path to getting there based on its own reasoning and decision-making. Agents can even work with other AI agents to form a whole team of digital coworkers who never need coffee breaks. Multiple agents working together is called an Agentic AI system.

Think of the difference between GenAI and Agentic AI this way:

>> **GenAI is like a calculator.** You push the buttons, and it gives you an answer.

>> **Agentic AI is like a junior assistant.** You tell it what outcome you want, and it figures out the steps to get there. Sometimes, it asks other human or AI assistants for help along the way.

Welcome to the age of Agentic AI. Buckle up. It's going to be an interesting ride!

About This Book

The shift from AI output (GenAI) to AI action (Agentic AI) is a huge technological feat on par with the autonomous cars available now, and these technologies share many similar risks and opportunities. Agentic AI can schedule tasks, run experiments, optimize business decisions, or help you shop online, for example. And it can do all of that without you having to hold its hand every step of the way.

But the addition of partial or complete autonomy that Agentic AI possesses comes with a new set of challenges. In an accounting scenario, how do you keep agents from going off track and deleting an entire spreadsheet or database? In a medical situation, how do you trust Agentic AI to work safely and not hurt a patient? If you love the idea of an AI agent working as your personal online shopper, how do you keep that agent from buying things with your credit card that you didn't intend for it to buy? And how do you tell the difference between a useful agent and one that's more hype than help?

This book is here to answer some of the practical questions about Agentic AI and explain the scope of applications that it can touch. By the time that you finish reading this book, you'll be able to talk about Agentic AI with confidence, spot it when you see it, and know how to make it work for you, instead of the other way around.

Some typical conventions that you may find in this book include the following:

>> If you see a term in *italics*, you can usually find a definition or explanation for the term close by in the text.

>> Web addresses and programming code appear in monofont. If you're reading a digital version of this book on a device connected to the internet, you can click the web address to visit that website, like this: `www.dummies.com`. Some web addresses break across two lines of text. If you're reading this book in print and want to visit one of these web pages, simply type the address exactly as it appears in the text, pretending the line break doesn't exist.

To make the content more accessible, I divide it into five parts:

>> **Part 1: Understanding Agentic AI:** In this part, you find out what Agentic AI is and how it works

>> **Part 2: Getting Started on the Agentic AI Path:** Check out this part to get a good grasp on where and when to use Agentic AI as well as the ethics involved

>> **Part 3: Agentic AI in the Real World:** Here you'll find how Agentic AI will likely change your work and your world — and what you can do to make sure no one is harmed in the process

>> **Part 4: Exploring Myths and Realities:** Here are the facts about what Agentic AI is and isn't, whether it has agency, and what amount of upskilling you need to keep pace

>> **Part 5: The Part of Tens:** This section gets right to the point of unexpected surprises now and 10 years from now, and it gives you a list of things that Agentic AI is absolutely terrible at doing

Foolish Assumptions

I wrote this book for anyone who wants to understand and use AI agents and Agentic AI systems in their work and daily life, as well as to prepare for inevitable changes that this technology will introduce. This book is written for professionals, not programmers. To get value from this book, you do *not* need

>> A degree in computer science, math, or engineering

>> Years of coding experience

>> A deep knowledge of AI research papers and technical protocols

If you can read, think critically, and apply new ideas in your work, you already have the background that you need.

But I do make certain assumptions about the book's audience (you) as a practical matter. For instance, I assume that

>> You possess at least a limited understanding of GenAI and are in hot pursuit of leveling up your skills to now understand and work with autonomous AI agents.

>> You have at least a basic level of comfort and skill in working with computing devices, browsers, and web applications.

>> You're smart and pressed for time, so you want all meat and no fluff in a fast and easy read. (I hope I hit that mark for you with this *For Dummies* book.)

Icons Used in This Book

Throughout this book, icons in the margins highlight certain types of valuable information that call out for your attention. Here are the icons that you might encounter and a brief description of each.

TIP

The Tip icon marks tips and shortcuts that you can use to make building, tasking, or using AI agents easier or simply more fun.

REMEMBER

Remember icons mark the information that's especially important to know. To siphon off the most important information in each chapter, just skim through until you find these icons.

TECHNICAL STUFF

The Technical Stuff icon marks information of a highly technical nature that you can normally skip over. Unless, of course, you came for the technical stuff — in which case, it's now earmarked for you.

WARNING

This icon warns you of a stumbling block or danger that may not be obvious to you until it's too late. Please make careful note of warnings.

Beyond the Book

In addition to the abundance of information and guidance related to Agentic AI in this book, you get access to even more help and information online. Check out this book's online Cheat Sheet: Just go to www.dummies.com and enter "Agentic AI For Dummies Cheat Sheet" in the Search text box. Press Enter, and a link to the Cheat Sheet pops up in the results.

Where to Go from Here

This is a reference book, so you don't have to read it cover to cover (unless you want to soak in all the new information all at once). Also, feel free to read the chapters in any order. Each chapter is designed to stand alone, meaning you don't have to know the material in preceding chapters to understand the chapter that you're reading. Start anywhere and finish when you feel you have all the information that you need for whatever task you have on hand.

Here are a few specific tips on where to find the info particularly interesting or useful to you:

>> Check out the Table of Contents at the front of the book or the Index at the back to find a topic of interest.

>> If you simply want to understand what AI agents are and how they work, both alone and together, read Part 1.

>> If you have business interests, Chapter 5 guides you through making a plan so that your investments in Agentic AI can deliver, both on the mission that you give it and a bankable return on investment. And Chapter 6 offers a peek into how first-adopter companies and industries are using Agentic AI at the time of writing.

» Chapter 7 poses all the hard questions that no one wants to grapple with —
but also that no one can escape. Here, I blow away the hype and present the
facts and obstacles that keep this tech from being a plug-and-play miracle.

» The chapters in Part 3 give you a good look at how Agentic AI technology is
reshaping work, economies, and safety for humankind.

» Do you wonder if and when people should think about AI having agency
and autonomy? Chapter 13 explores the issues of consciousness, intent,
and motive as it applies to AI agency in sharper detail.

1

Understanding Agentic AI

IN THIS CHAPTER

» Identifying Agentic AI and its connection to traditional AI

» Embracing reasoning as an Agentic AI trait

» Distinguishing AI agents from Agentic AI

» Evolving prompting to direct Agentic AI

» Examining Agentic AI impacts on the internet and commerce

Chapter **1**

Introducing Agentic AI

A gentic AI represents a significant shift in the evolution of AI. Its capabilities are leaps and bounds beyond those of Generative AI (GenAI) and other traditional forms of AI, such as voice assistants like Siri and Alexa, or the technologies that drive autonomous cars. The most distinguishing feature that puts Agentic AI in a league of its own is *autonomy* (its capacity to make decisions and carry out a set of actions toward a goal without specific instruction at each step).

Put in a simpler way, GenAI is all talk, and Agentic AI is all action.

This chapter explores what defines an AI as agentic, how it differs from other AI types, and the foundational elements required to recognize such a system. Further, this chapter presents two examples of profound disruptions that Agentic AI brings: the rise of the AI web and the shift from e-commerce to A-commerce.

Defining Agentic AI

Agentic AI is a type of artificial intelligence that can act on its own to achieve goals, instead of just waiting for prompts from a human. It doesn't just respond to a human's commands at every step. It can decide what steps to take, make plans, change its approach if something doesn't work, keep track of what it learns along the way, and reflect on its performance so that it can improve.

The word *agentic* comes from *agent,* meaning that the AI behaves like an agent on your behalf, an intelligent helper, or a problem-solver that has a degree of independence. Think of Agentic AI as AI that not only completes tasks, but also figures out how to complete them.

This isn't the AI of scary science fiction stories. However, technology experts widely view Agentic AI as a critical stepping stone on the path toward artificial general intelligence (AGI; the form of AI depicted in scary science fiction stories) and possibly the *technological singularity* — or simply *the singularity* — a hypothetical future point at which artificial intelligence surpasses human intelligence in a way that leads to unpredictable, rapid, and irreversible changes in society and technology.

Moving toward AGI with Agentic AI

By enabling systems to reason, plan, reflect, and take initiative across changing environments, Agentic AI helps bridge the gap between today's highly specialized models and the broad, self-directed intelligence that systems need to realize AGI.

In the broader vision of the hypothetical singularity, exponentially advancing artificial intelligence could emerge through self-improving, interconnected agentic systems. Such systems might continually refine their own architectures and methods, collaborate with other agents (even across different networks or domains), and pursue complex goals with diminishing need for direct human oversight.

However, Agentic AI and more autonomous systems bring us a step closer to the kind of self-directed intelligence imagined in singularity scenarios, and they also introduce new risks of unintended consequences and demand additional safeguards, including

>> **Strong alignment with human values,** training and guiding models by using data, feedback, and objectives that reflect humanity's shared ethical and social principles.

>> **Robust guardrails** that provide clear, operational boundaries and fail-safes that define what the AI can and cannot do, even as it learns or acts independently.

>> **Ethical oversight** to maintain human accountability in how developers design, deploy, and monitor agentic systems throughout their lifecycle.

REMEMBER

Agentic AI is not only a technical milestone in the evolution toward AGI, but also a pivotal moment in which to decide what kind of AI future humanity will build.

Noting that agentic systems already exist

Developers often design Agentic AI to handle complex, multi-step processes, such as managing projects, conducting research, or solving technical problems. These systems can include tools such as memory (to track what's already been done), reasoning engines (to decide what makes the most sense), and planning modules (to map out steps and sequences). Although these systems aren't really human-like or conscious, they're moving closer to becoming trusted aides to the people using them.

TIP

Despite the advances made so far in the development of Agentic AI, it's an emerging technology; people will likely take some time before they accept this technology as a routine entity in their work and daily lives.

Reasoning as AI's Way Forward

Whether people consciously recognize it (or not), instinct often drives them to fear an AI's ability to reason independently. This fear stems from the understanding that the capacity to reason has long defined humanity's unique position at the top of the natural order.

If machines can also reason, the thinking goes, they could eventually become intellectually superior to humans, creating an unnatural hierarchy of reason. That perceived loss of uniqueness threatens the intrinsic value of being human and may also feel like a challenge to humanity's place as both the observer and steward of the natural world around them.

Exploring philosophy, reason, and fear

The instinctive fear of being knocked from nature's top spot is rooted deep in humanity's collective history. Across time, people have moved through this line of thinking:

REMEMBER

- ❯❯ **Recognizing the importance of reason:** Classical Greek philosophers introduced the idea that *reason* is the defining characteristic separating humans from all other creatures, and is the source of our unique position in the natural order of the world.

- ❯❯ **Using reason:** Reasoning rose to be the centerpiece of Western philosophy and later of Western science. Although Western thought also includes faith and empirical observation, reason still serves as its guiding foundation; the method by which truth and knowledge are pursued.

- ❯❯ **Defining humanity with reason:** Western philosophers view the ability to reason as both a noble pursuit of truth and a uniquely human capacity. Reasoning, as reason has it, is what makes humans — well, humans.

- ❯❯ **Expanding reasoning capabilities:** Because reason (philosophically) is held as a noble pursuit, many people feel driven to improve upon how, how fast, and how much they can reason — which leads to the creation of tools to extend human reasoning. These tools range from mathematics and logic to computing — and now AI.

I generalize and minimize the description of philosophy because this is a book about AI's ability to reason — or lack thereof — and not about philosophy. But I bring philosophy to your attention so that you can see the root of humanity's unease. AI threatens not just to imitate human reason but to redefine it. That's the source of both the fear and the fascination, and the devilishly desirable urge to build something that can reason as we do, perhaps even better.

Putting AI reasoning into perspective

In the beginning — which is to say, a few years ago, when ChatGPT became a household word, and not decades ago, when Generative AI (GenAI) became a thing — the public wasn't sure what to make of the tool. But thoughts immediately turned to the deduction that if AI is intelligent, then it must be like (or even better than) humans, and therefore both capable of reasoning and inherently evil. Right? Never mind that various levels of intelligence, cleverness, and reasoning exist, some of which barely blip on the scale between instinct and thinking.

Although creators and marketers of GenAI models have heavily promoted their systems' supposed reasoning abilities, many industry observers and technology journalists, including myself, have long expressed skepticism about these claims.

EXPERIENCING THE ILLUSION OF THINKING

In June 2025, Apple (a technology giant) published a research paper from its Machine Learning Research team titled (in short) "The Illusion of Thinking," which systematically evaluated large language models (LLMs) and large reasoning models (LRMs) by using controlled puzzle environments and specialized datasets. The findings felt like an earthquake under the feet of AI designers. The groundbreaking study revealed critical limits in the reasoning capabilities of today's AI systems. In short, the researchers found that LLMs and LRMs don't really "think." They simply recognize patterns and imitate the steps of reasoning rather than truly understanding what they're doing.

The takeaway: While developers now have access to many AI models and tools, none of them can truly reason in a humanlike way. Still, some developers are working to create Agentic AI systems that can reason on their own with little to no human intervention. Is the effort doomed from the start? No. But serious challenges remain. For example, AI models tend to break down when tasks become too complicated or require deep, step-by-step logic. Apple's report points to a few ways researchers might get past these hurdles, such as creating smarter test environments that actually measure real reasoning, and teaching AI to check its own work as it goes along.

The path to truly autonomous AI agents is still blocked by basic limitations (such as needing more high-quality, well-structured data and lacking true common sense). Apple's research findings (see the sidebar "Experiencing the Illusion of Thinking," in this chapter, and the section "Differentiating between AI Agents and Agentic AI," later in this chapter) highlight a critical issue: Even the most advanced AI systems today are great at mimicking thought but not at genuinely understanding. They often rely on techniques like chain-of-thought prompting (a method that walks the model through a problem one step at a time) to appear more thoughtful, but they're still missing the logical depth and common-sense reasoning needed for full independence.

TECHNICAL STUFF

When Apple researchers talk about a *reasoning collapse,* they mean that AI models can look brilliant on simple tasks but start to fall apart as problems get harder. The models don't understand logic, but instead, they follow patterns they've seen before. Once those patterns no longer fit, their "thinking" breaks down.

TIP

Don't confuse reasoning behavior with reasoning ability. When an AI explains its steps or thinks aloud, it's not actually reasoning. It's replaying patterns from training data that look like reasoning. True reasoning means understanding cause and effect, spotting breaks in logic, and adapting on the fly. These are things today's AI still can't do (at least not yet).

Recognizing the operational challenges of Agentic AI

Beyond reasoning limitations, Agentic AI systems face substantial technical and operational challenges that complicate real-world deployment. Agentic AI archi-tectures are complex and require robust infrastructure, sound data governance, and seamless interoperability with existing systems to function at scale. (Flip to Chapter 3 for more about the components that Agentic AI requires.)

Orchestrating multi-agent workflows, maintaining accuracy, and managing vari-ability remain major technical hurdles. For Agentic AI systems to reach their full potential, designers and developers must overcome both reasoning gaps and practical constraints. Anything short of genuine autonomy is likely to appear as a connected series of specialized (but conventional) AI models, coordinated to per-form tasks together rather than truly acting on their own.

Differentiating between AI Agents and Agentic AI

The two terms *AI agents* and *Agentic AI systems* are often used interchangeably in discussions about artificial intelligence, but they actually describe different con-cepts that share overlapping features. Understanding their differences can help clarify the different types of AI that exist today and where they might be headed as they evolve toward more capable and autonomous systems:

>> **AI agents:** Software entities designed to perform specific tasks autonomously within defined parameters. They operate based on programmed rules (rule-based agents), machine learning models (learning-based agents), or large language models (LLM-based agents) to achieve particular objectives by perceiving their environment, making decisions, and taking actions. Examples include customer-service chatbots, game-playing bots, web navigation agents, and recommendation systems.

>> **Agentic AI systems:** Also software entities, but they extend basic task automation into the realm of complex, multi-step process management. These systems can plan their own actions, coordinate multiple tools or agents, and adapt their workflows to reach broader objectives, often across less predictable environments.

Apple 2025 research study, "The Illusion of Thinking: Understanding the Strengths and Limitations of Reasoning Models via the Lens of Problem Complexity" (which you can read about in the sidebar "Experiencing the Illusion of Thinking," in this chapter), focuses primarily on large language models (LLMs) and large reasoning models (LRMs), and not explicitly on Agentic AI. However, because Agentic AI systems are typically built on top of similar LLM foundations, the study's findings apply by extension. The research confirms that both AI agents and, by association, Agentic AI systems rely heavily on pattern matching, rather than genuine reasoning. This dependence limits their performance on tasks that demand abstract, logical, or novel reasoning.

TECHNICAL STUFF

For Agentic AI systems, Apple's study points toward several potential directions for improvement. These include hybrid neurosymbolic architectures (which combine neural networks' pattern-recognition strengths with symbolic reasoning's logic and structure) and the use of advanced datasets such as GSM-Symbolic to better evaluate and train reasoning skills. The study also introduces Twisted Sequential Monte Carlo (TSMC), a technique for improving multi-step reasoning and inference. It is an approach that could be particularly valuable for agentic systems striving for higher autonomy, whether that autonomy is designed in or gradually self-emerges through adaptation.

Meeting examples of agents and agentic systems

Examples of AI agents include chatbots such as ChatGPT (www.chatgpt.com; see Figure 1-1), recommendation systems that suggest what to watch or buy, or simple robotic agents, such as a Roomba vacuum cleaner. These agents typically have a narrow scope and focus on well-defined tasks with limited adaptability to new contexts.

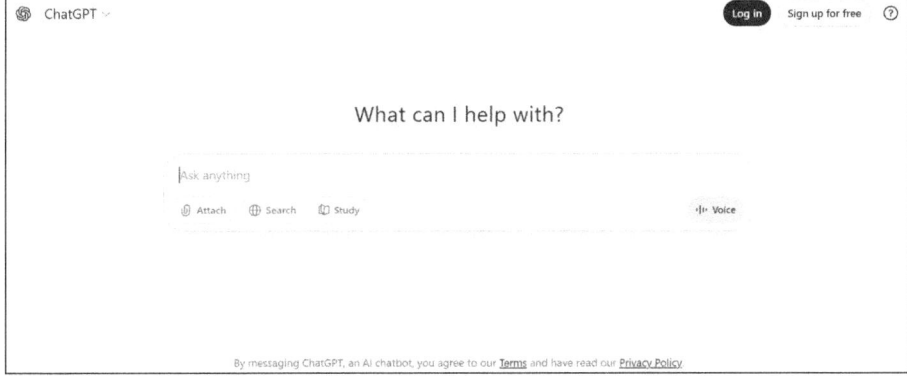

FIGURE 1-1: A screenshot of ChatGPT chatbot user interface.

www.chatgpt.com

An Agentic AI system goes further than a chatbot such as ChatGPT. Agentic systems often combine multiple AI agents and add capabilities such as goal-setting, planning, reasoning, and task monitoring over extended periods. One example is Godmode (`www.godmode.space`), shown in Figure 1-2, an online interface that lets you launch and manage autonomous AI agents such as AutoGPT (`www.agpt.co`) and BabyAGI (`www.babyagi.org`). Godmode doesn't crowdsource from these projects; instead, it provides a user-friendly control panel that connects to and coordinates open-source agent frameworks behind the scenes.

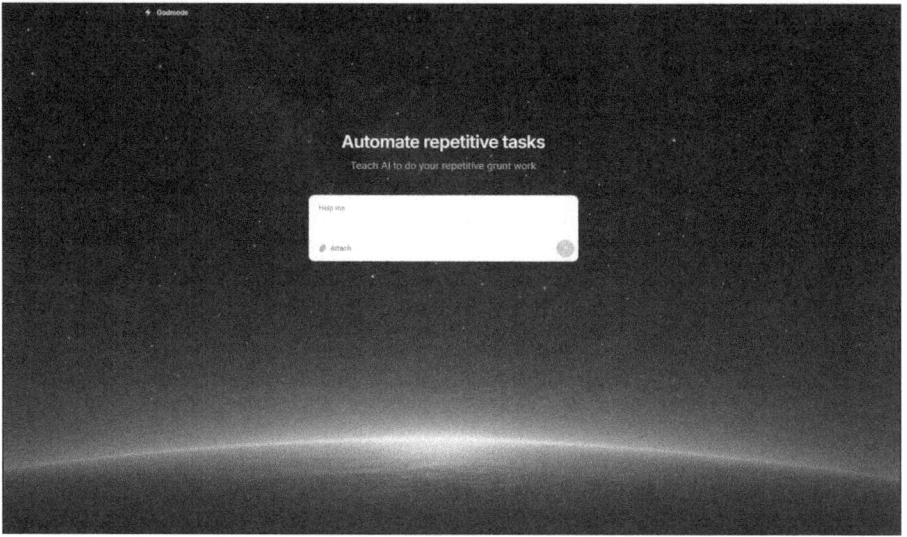

FIGURE 1-2: A screenshot of the Godmode interface.

`www.godmode.space`

If you'd like to try these systems directly, you can access a hosted version of BabyAGI in your browser (with no local setup needed) at `https://babyagi-ui.vercel.app`. For open-source enthusiasts, both AutoGPT and BabyAGI also offer their own graphical user interfaces (GUIs) through GitHub:

>> **AutoGPT UI:** `https://github.com/neuronic-ai/autogpt-ui`

>> **BabyAGI UI:** `https://github.com/miurla/babyagi-ui`

Other examples of Agentic AI systems include autonomous supply chain management platforms that optimize logistics in real time, and AI-driven research assistants that design experiments, gather data, and summarize findings. Another emerging example is GPTConsole (`www.gptconsole.ai`), which uses autonomous agents such as Pixie to generate complete, production-ready apps and websites from simple landing pages to full data dashboards. Figure 1-3 shows the free landing-page generator interface that you can try yourself.

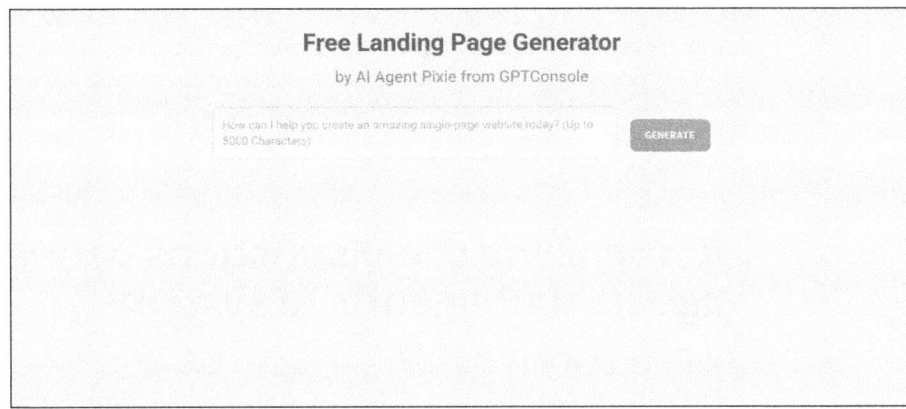

FIGURE 1-3:
A screenshot of an AI agent pixie that offers a free landing-page generator.

Seeing what agents and Agentic AI systems have in common

Although I define AI agents and Agentic AI systems earlier in this chapter (see the section "Differentiating between AI Agents and Agentic AI"), it's worth restating for clarity's sake that the term *AI agent* is broader than either Agentic AI or GenAI tools. Using the term doesn't automatically imply that an AI agent is agentic or generative when in fact it could be neither but instead a different type of AI altogether. In this book, AI agent means any software entity that can perceive its environment, make decisions, and take actions toward a goal, with or without direct human control.

Both AI agents and Agentic AI systems operate with some degree of autonomy so that they can make at least some decisions without constant human intervention. For instance, a chatbot (AI agent) responds to user queries independently, while an Agentic AI system that manages a logistics network adjusts delivery truck routes in direct response to traffic data, weather conditions, road conditions, and/or real-time supply updates.

By design, both AI agents and Agentic AI systems can

>> **Perceive and interact with their environments.** A recommendation system (an AI agent) analyzes user behavior and preferences, while an Agentic AI system in healthcare might monitor patient data and adjust treatment plans based on vital signs, lab results, or medication responses.

>> **Achieve specific objectives.** A virtual assistant (an AI agent) aims to answer questions accurately, while an Agentic AI system for urban planning seeks to optimize traffic flow across an entire city.

>> **Leverage AI technologies,** such as machine learning, natural language processing (NLP), and/or rule-based systems. Large language models (LLMs), for example, can power both a conversational AI agent and components of an Agentic AI system.

Recognizing the distinctions between agents and Agentic AI systems

Although AI agents and Agentic AI systems have similarities (see the preceding section), you can find marked differences, as well. Here are some examples:

>> **Their strengths:**

- *AI agents:* Offer simplicity and efficiency because they're highly optimized for repetitive, well-defined tasks, and deliver fast response times. They're also cost-effective and generally reliable in their narrow domains and tasks.

- *Agentic AI systems:* Excel in adaptability and scalability in handling complex, multi-step problems that require coordination across multiple agents and tools. They are also showing emerging potential for advanced reasoning and collective decision-making.

>> **Their weaknesses:**

- *AI agents:* Have limited scope of data and functionality, *reasoning deficits* (they may fail to grasp the full intent of a prompt), *brittle performance* (producing nonsensical results in complex real-world situations), and low capacity for *generalization* (applying training to new data).

- *Agentic AI systems:* Present complexity for developers and users, reasoning limitations (because of limitations of available data and training), high development and operational costs, and their still-experimental nature.

>> **Their ideal use cases:**

- *AI agents:* Work well as customer support chatbots, shopping and other recommendation systems, simple automation (of data entry or appointment scheduling, for example), and robotic process automation (RPA).

- *Agentic AI systems:* Can optimize supply chains (to increase efficiency and reduce costs), support healthcare decision-making (providing patient data analysis and actionable insights for treatment), do scientific research (for handling large datasets and predicting outcomes), and manage smart cities (including optimizing transportation services and urban planning).

Mapping the Path from Prompt Engineering to AI Autonomy

The path from prompt engineering to AI autonomy doesn't involve outright replacement of one by the other, but an evolving relationship (which I talk about in Chapter 4). You'll still need foundational prompting skills to work with increasingly intelligent systems. While AI becomes more *agentic* (that is, autonomous), the role of human input shifts in its focus, but not in its value.

TIP

In my opinion, people who now claim that AI users don't need to learn how to prompt the systems that they use are (at best) foolishly misguided. In my previous book, *Generative AI For Dummies* (Wiley), I discuss how this technology enables machines to speak like humans, but to effectively use them, humans must think like machines. By that, I mean people who use AI need to think in a logical, step-by-step progression, in the same way that developers do when they write computer programs. Prompting skills teach you how to think like a developer.

Further, prompts will remain the primary interface for you (and other humans) to communicate with AI agents — especially in Agentic AI systems. The reason is simple: You can more easily command the AI in your own language than in computer code; then a large language model (LLM) understands your command and parses the necessary instructions to different AI agents to perform the feat that you commanded. Chapter 4 goes into detail about how to interact with Agentic AI.

REMEMBER

Even advanced agentic systems use prompting under the hood to pass instructions, context, and goals between modules or agents. Despite their complexity and apparent autonomy, much of the system's internal coordination still happens through prompt-based messages. And prompting isn't the only way agents communicate. Some frameworks also use function or API calls, shared memory, or symbolic reasoning layers to exchange data and results.

Simply put, prompting isn't just for humans' use when talking to AI; it's one of the key ways that AI components use to talk to each other inside modern Agentic AI systems.

Engineering your prompts for successful AI interaction

Prompt engineering involves crafting precise, contextual instructions for LLM-based agents and tools so that you get the results you want. You can use it in all kinds of applications, from content generation and coding, to research assistance, data transformation, and even tool automation. Prompt engineering teaches you how to use structured language to interact with and direct powerful GenAI systems. Think of good prompting like giving instructions to a very smart assistant who understands your language but still needs clarity and direction.

As developers began experimenting with more complex interactions, they started chaining prompts — using the output from one prompt as the input for the next. Frameworks such as LangChain and AutoGPT made this easier by allowing AI systems to simulate multi-step reasoning or tool use. Chaining prompts connect actions such as using memory, making decisions, and calling application programming interfaces (APIs) or plugins. This approach moves AI from answering one-off queries toward executing full workflows. In that sense, prompt engineering at this level is less like writing a single question and more like designing a system or scripting a process. It is the creative foundation of Agentic AI behavior.

For example, an Agentic AI system for customer support might use a prompt *pipeline* (chain) to classify queries, retrieve relevant data, and generate responses, with prompts ensuring tone, context, and accuracy. Although Agentic AI systems can set sub-goals, evaluate their own outputs, and operate in open-ended, dynamic environments, they still rely heavily on well-crafted prompt templates, safety guardrails, and human review cycles. Even autonomous agents are largely built atop *prompt scaffolding*, which is a prompt engineering technique that uses a series of prompts to guide an AI system.

REMEMBER

Even AI that developers see as autonomous has no will, no self-awareness, and no inherent motivation to do anything. Humans must give AI a reason to act, even the over-industrious Agentic AI systems. Without a call to action and a cause to complete, any AI will sit idle for centuries.

Controlling AI system actions with effective prompting

Prompting gives you a way to guide and influence unpredictable systems and perhaps stop unsafe actions (which you can read more about in Chapter 7). Effective prompting can reduce hallucinations, align outputs with your intent, and promote safer, more ethical system behavior. As AI agents grow in autonomy, troubleshooting their behavior often involves analyzing their internal prompts and instruction chains to understand what went wrong. Users must have the necessary prompting skills to do this kind of work. Think of it as AI quality assurance or cognitive debugging.

In enterprise environments, prompt engineering goes even further. Organizations often build custom prompt libraries, design domain-specific prompts, and develop fine-tuned instruction sets for areas such as law, medicine, or customer service.

These libraries are growing quickly because well-crafted prompts are often more scalable and cost-effective than retraining or fine-tuning models from scratch.

In short, the ability to frame questions, shape outputs, and design workflows through prompts is a creative and strategic skill, not just a technical one, which remains critical in human-AI collaboration, no matter how autonomous AI becomes.

Viewing prompting as foundational

Ultimately, prompting acts as a core design layer in AI interaction and development. In the same way that knowing SQL doesn't become irrelevant when databases get more complex, prompt engineering won't disappear while AI becomes more autonomous. It will simply evolve into AI literacy and collaborative creativity for individual AI users.

At the maker and enterprise levels, prompt engineers will likely evolve into AI system architects, agent behavior designers, and human-AI interaction specialists. The essential skill of translating human intentions into AI-understandable instructions remains fundamental, even as it operates at higher levels of abstraction and complexity. Chapter 8 dives into the potential changes to how people will work when they have Agentic AI in the picture.

The Budding Agentic AI Web

Today, the internet mostly reacts to what people ask it to do. You click a link, fill out a form, or type a question, and something responds. But in the near future, I predict that AI agents will start to do these things on their own — behind the scenes — based on what they know about your goals, habits, desires, and preferences.

The idea of the Agentic AI web is the next big shift in how people use the internet. Essentially, the internet will transform into an environment where people rarely visit but where smart AI assistants work together and get things done for people. You can see that beginning to happen now as more people use AI rather than search engines to find information and rarely bother to go to the website or original source. Chapter 9 looks at the role Agentic AI will play in the online marketplace.

Expanding the duty of agents online

In the future, the AI agents won't just live in one app or service; they'll connect across websites, platforms, and tools. One agent might handle your schedule, another could manage your online shopping by acting as your personal shopper, and another might track job openings or negotiate a better deal with your ISP or streaming service on your behalf. The various agents can talk to each other, use tools, search the web, write e-mails, fill out forms, and even call APIs. They'll take action largely on their own initiative and without needing you to hover over them.

So you won't go to the internet to watch videos, listen to music, shop, or do research. Instead of going to websites directly, you'll rely on a network of smart helpers working quietly in the background. Instead of having to log in to ten different websites to get something done, you might just tell your agent what you want to do (for example, compare the prices of an item offered by three vendors), and it takes care of the rest. You're still in charge, but the agents handle the details for you.

For example, suppose you ask an AI assistant to plan a trip. As part of an Agentic AI system, the assistant could pass tasks to other AI agents — one agent books flights, another checks traffic, and a third picks activities that it thinks you'd love to do. The AI agents work together smoothly to get the job done without further input from you.

Scaling up to a citywide, nationwide, and global reach

A broader span of agent authority in the future could also scale up to big stuff, such as Agentic AI systems running cities, managing energy, or speeding up the impact of science by sharing discoveries globally. These AI agents could learn on their own, teaming up for large tasks, such as disaster relief or prevention perhaps by using drone data and weather updates to save lives from a disastrous storm.

For the internet to morph into this AI-run version of itself, it needs a few key things to come together:

>> **A shared language that enables agents and systems to talk to each other:** Several emerging standards already work toward that goal and one of the most promising is the Model Context Protocol (MCP). MCP acts like a USB-C cable for AI: a universal connector that allows different AI systems to plug into tools, apps, and data sources without needing a custom setup each time. It solves the problem of one-off integrations by creating a standard way for AI to access external resources such as files, databases, or web services.

>> **Secure ways to share knowledge between systems and agents:** For example, healthcare systems (such as hospitals and clinics) need to swap patient data, treatment plans, or medical advice without exposing private information. An AI system working in this environment must manage multiple tasks simultaneously, maintain strict data security, and comply with privacy regulations such as HIPAA. To support this, some developers explore using trust-enhancing technologies, including blockchain or other tamper-evident record systems, to securely log transactions, verify data sources, and track information access across agents and systems.

The key components of an AI-run internet are still in their early days, but developers are addressing the issues of shared language and secure data while increasingly setting the stage for huge advances in autonomous AI operation.

Following the Shift to A-Commerce

You probably won't just get up one day to find that Agentic AI systems have completely restructured the internet into some alien digital landscape. Instead, you'll likely note subtle changes in your online interactions first, followed by bigger leaps of disruption. Indeed, you likely noticed a few changes online already, especially in how you search for and consume information.

Online search used to be simple: You typed a few keywords into a search bar, got a list of links to source websites in return, and then had to dig through websites and pages to find what you needed. GenAI tools such as ChatGPT, Google's Gemini, Microsoft's Copilot in Bing, and Perplexity AI (the first major AI-native search engine that doesn't rely on results from traditional providers) have changed that experience dramatically.

Now, instead of just returning links to source websites and a few paid ads, search engines serve up concise answers from AI responses. Today, you can commonly get conversational, direct answers to your query from an AI tool with nary a source cited.

Recognizing the fallibility of AI answers

Having AI provide concise answers can be dangerous because many users think all AI results are dependable, when in truth, they're not. Even Perplexity AI, which routinely serves up sources with its narrative results, can lead you astray. Any AI can make up sources as easily as it can make up facts. Yes, I mean AI can outright lie to you about who said what and where and when. Source citations often aren't worth the pixels used to display them.

Any AI outputs, even the summaries given on search engine results, need to be fact-checked before you accept and use them. First, click through the source links the AI provides (if it does provide them) to ensure the source exists and that it is the original source of that particular information. You can also scroll down and use the search engine results that appear below the AI summary to find reputable sources to peruse so that you can validate and verify the answer that the AI gave you as a summary.

All told, that doesn't provide much of a time saver or a convenient service, now does it? Still, it makes a step in the direction of A-commerce.

Reducing specific website interaction

Online search results are changing fast. Even traditional search engines now place AI-generated summaries at the top of the page, ahead of the familiar list of website links that formerly defined search. The reason is simple: Competition for use and user trust is fierce. Tech companies are racing to make their AI systems the first stop for everything people want to know. As users increasingly accept these instant summaries instead of clicking through to source websites, overall web traffic is dropping across the board.

This shift can drastically limit how many page viewers a news article gets, how many online sales a merchant makes, how many students a university attracts, and so on. In other words, e-commerce and content-driven industries alike suffer in innumerable ways as AI tools become the new intermediaries between information and its audience.

Adding insult to injury, while AI-powered summaries gather more eyeballs, the classic search engine optimization (SEO) strategy of ranking high on a results page matters less if users no longer click through to websites. If organizations can't improve their website's visibility and number of click-throughs via SEO and search result ad strategies, then why bother having a website? And how can those organizations survive and prosper without direct traffic?

The tipping point will come when AI systems, not human users, handle most product discovery and purchasing decisions online. That moment marks the shift from e-commerce (electronic commerce) to A-commerce (short for autonomous commerce) where AI agents search, compare, negotiate, and even complete transactions on behalf of users.

Relying on personal AI shoppers

Although e-commerce sees people browsing websites, comparing prices, and clicking a Buy button, A-commerce flips that model completely. In the A-commerce scenario, AI agents will act as your personal shoppers, deal hunters, and decision-makers by handling the shopping process for you based on your goals, desires, preferences, and budget. Agents may even purchase the items for you — using your credit card, of course — and have the item sent to you. All of that shopping and buying may happen with you doing little more than expressing an interest or intent to buy an item or service.

To better help you imagine how this AI shopping agent will work, consider the following examples:

>> **You need a new pair of running shoes.** Your agent knows your size, preferred brands, past purchases, and current deals across the web. It finds three great options, checks reviews, and presents you with the best one — or just buys it if you previously gave it permission.

>> **You're planning a trip.** Instead of jumping between flight, hotel, and rental car sites, your travel agent AI bundles the whole thing, rebooks if prices drop, and gives you one neat itinerary.

>> **You run a small business.** Your AI handles supplier orders, monitors inventory, and even renegotiates contracts when better options appear.

>> **You have an important event on your calendar.** Your AI notes the event and appropriate attire, and searches for the best deals on each item of the perfect outfit — from shoes to hat. All of the outfit fits your size and list of preferences. The AI can then ask you to choose between outfits or items, and then purchase it. Or, if it has your permission, simply buy it and have it shipped to your home or the hotel where you'll be staying during the event.

In short, A-commerce is the next evolution in online commerce, where AI agents handle purchasing decisions, product research, and transaction completion on behalf of you, your family, or your company. This shift will create profound changes in how SEO strategies and websites must evolve.

Optimizing commerce sites for AI

Traditional SEO and websites target traditional search engine rankings to increase visibility before people who are interested in buying their products or services, but A-commerce requires optimizing for AI agents that make purchasing decisions without a person looking at any of their wares. So site designers need to structure content for AI comprehension, rather than solely for human readability.

Sites need to present product information in formats that AI agents can easily parse, compare, and evaluate.

Consider these factors that can reshape the content of commerce-related websites:

>> **Structure and focus of website content:** AI agents rely heavily on structured data to make decisions. For example

- E-commerce sites will need to prioritize schema markup, product APIs, and machine-readable formats over traditional content optimization. The focus shifts from keyword density for making the website visible to human searchers to data clarity and accessibility for AI agents.

- AI agents will evaluate different trust signals than human consumers. Things such as API reliability, data accuracy, real-time inventory status, and transaction security will become more important than traditional brand signals. Emotional and impulse buying will likely fall to the wayside unless merchants can find a way to get the AI to consider more than its precise objective.

- Traditional SEO focuses on attracting human visitors to websites, but A-commerce SEO focuses on helping AI agents find information and complete transactions. That means web designers must prioritize machine-readable, conversion-friendly signals — things AI agents can easily recognize and act on, such as clear pricing, real-time availability, shipping details, and return policies.

>> **Variety of inputs and interactions:** A-commerce AI agents will process text, images, video, and audio to make purchasing decisions. SEO strategies must optimize across all these formats, ensuring product information is accessible, regardless of how the AI agent consumes data. For example, an AI shopping assistant might read a product description, verify price tags embedded in an image, watch a short video to assess features, and even analyze customer-review audio clips for sentiment — all before recommending or completing a purchase on your behalf.

- A-commerce SEO will need to be optimized for voice- and chat-based interactions, where AI agents gather product information through conversational queries. This optimization requires content that answers specific questions that AI agents ask when comparing products or evaluating purchases.

- You and other users may even make custom stipulations for their AI assistants, for example, by excluding suppliers based on political affiliations, DEI policies, or sustainability practices. Understanding how different AI agents make purchasing decisions will become a core SEO competency.

>> **Updating content in real-time:** A-commerce operates at machine speed, requiring SEO strategies that can adapt instantly to changing inventory, pricing, and demand patterns. Traditional static optimization approaches (such as periodic keyword tuning or quarterly content refreshes) will become less effective because AI agents will expect current, verifiable information for decision-making. AI agents may even ignore or down-rank websites, vendors, or service providers if their data appears out-of-date by even a few minutes, especially in fast-moving markets like travel, logistics, or retail.

Navigating the transition to A-commerce

Ultimately, the transition to A-commerce won't eliminate SEO, but it will transform it into a more technical, data-driven discipline, focused on AI agent behavior, rather than human search patterns. Success will require understanding both traditional search optimization and the decision-making processes of AI systems that increasingly mediate between consumers and commerce platforms.

If A-commerce takes off, it may lead to a dramatic drop in a person's screen time. If your agents are doing the searching, comparing, and transacting for you, you have to step in only when decisions matter or exceptions arise. You'll no longer have to deal with endless tabs, pop-up ads, or e-mail floods from retailers. Your attention can become more focused and your time freer because you won't need to go online to shop for or book something — your AI agent is already online 24/7, quietly optimizing your life in the background.

Over time, the widespread use of AI agents could radically reduce people's daily dependence on screens. Kids may no longer learn to surf the web as we know it; they'll grow up learning how to manage and supervise AI agents instead. (Check out Chapter 14 for more ways that your life may change thanks to Agentic AI.)

Chapter **2**

Peeking Inside the AI Agent Mind

This chapter gives you an overview of how agentic artificial intelligence (AI) operates today. You can find out how both standalone agents and a larger system of interacting agents go about their tasks. You may need a while to fully grasp the implications of an AI mechanism that performs beyond the level of following static instructions from a prompt. For now, you need to understand that Agentic AI is a giant leap in the ongoing AI evolution. This jump is like moving from riding horses as the primary mode of transportation to space travel in a single leap.

AI developers and engineers create Agentic AI to act autonomously in its pursuit of goals, decision-making, and active adaptations on the fly. Its behavior is based on changing information and context. A single agent might handle a complex task from start to finish, while in a multi-agent system, several agents coordinate with one another, each specializing in parts of a broader workflow.

In this chapter, you can discover how these AI agents plan their actions, communicate with each other, and execute tasks. You can also get a comparison of the capabilities of Generative AI (GenAI) and Agentic AI, and a better understanding of the mechanics behind this rapidly evolving technology.

Linking the Fundamental Building Blocks

AI developers and engineers construct Agentic AI systems from interconnected components that work together. But even with all this advanced automation, Agentic AI can't function alone. Human planning and oversight create the foundation not only in the developmental and task assignment phases, but throughout the Agentic AI's existence. Developers design and build agent workflows, define roles, create feedback loops, choose the right tools, and set safety boundaries. They also determine how agents access knowledge and when to involve humans in the loop.

Identifying Agentic AI building blocks

The Agentic AI system's core includes several key building blocks that collectively enable AI agents to operate with purpose and autonomy. These building blocks include

- ❯❯ **A mission or objective:** Each agent receives a clear objective from a human or derived from a broader task. The objective guides the agents and the overall system's actions. Each agent uses planning and reasoning to achieve its objective by breaking the task into smaller steps, prioritizing them, and determining the best course of action based on available resources.

  When the agent encounters complex challenges that require resolution to meet the objective, the agent uses a process called *task decomposition* to map out a series of simpler, more manageable subtasks. This process isn't unique to Agentic AI and commonly takes place in project management activities, software development, and robotics. People also use task decomposition in their personal productivity efforts.

- ❯❯ **Short-term memory and long-term memory:** The agent uses both memory types to recall information that's relevant to its current mission:

  - ● *Short-term memory:* Helps the agent keep track of recent actions. *Short-term* typically refers to information held within the current conversation or task session.

  - ● *Long-term memory:* Serves to improve performance, avoid repetition, and build contextual awareness over time. Long-term memory is essential for learning — for both humans and machines. *Long-term* refers to information stored persistently between sessions in mechanisms such as databases or vector stores.

>> Tool linkage and use: Agentic AI can't and doesn't act in isolation. It must interact with software tools, application programming interfaces (APIs), databases, and even other AI models to gather information, execute actions, and generate outputs. Technologies such as LangChain, AutoGen, and OpenAI's function calling help to link agents with external tools:

- *LangChain* helps AI agents interact with external data by using chains of reasoning. In this context, "chains of reasoning" refers to the framework's core concept of chaining together multiple steps or components to accomplish complex tasks that require sequential processing and decision-making.

  For example, in a use case, an agent can use LangChain to pull real-time stock prices from a financial API, process that information, analyze it, and summarize the results for human consumption.

- *AutoGen* allows multiple AI agents to collaborate and call tools or APIs, as needed. For example, when using AutoGen, one agent in the system can write computer code, while another runs it and returns with the output.

- *OpenAI's function calling* lets generative pre-trained transformer (GPT) AI models trigger external tools by defining functions in the code. A weather chatbot can use function calling to access and retrieve live data from a weather API, for instance.

New protocols such as Model Context Protocol (MCP), Agent Network Protocol (ANP), Agent-to-Agent (A2A) protocol, and Agent Communication Protocol (ACP) also help agents interact with other AIs, data sources, and software. You can find out more about these protocols in Chapter 3.

REMEMBER

World modeling offers another way in which AI agents can connect and use resources in more sophisticated ways. World modeling gives an agent the ability to build an internal representation of the environment that it's working in. Think of this capability as the ultimate in providing context to aid AI's understanding of the many nuances, conditions, and restrictions inherent to any given environment. And the agent can get far more context from working in a digital workspace, a data ecosystem, or any scenario that exactly replicates its existence in the real world. I talk more about world models in Chapter 3.

>> Communication and coordination: Agents need to share updates, delegate subtasks, resolve conflicts, and sync progress within themselves and with other agents in the system. Agentic AI systems achieve this sharing by using frameworks that enable messaging, state management, and collaborative decision-making, among other capabilities.

Enter the overseers

Without good planning and oversight, agentic systems can fail, suffer from model drift, or even create unsafe and disastrous events in the real world. After all, these systems aren't like yesterday's chatbots, prompted to perform on command and on task. These AI agents act largely on their own initiative.

Before you panic over AI agent autonomy, think about horses that are trained to perform tasks, from arena acts to tourist trail riding and field plowing. Horses typically do perform as expected, but when they don't, a rider or the horse itself can be seriously harmed. Additionally, if the horse doesn't have the right tools, such as a bridle, reins, or a plow, the horse either won't or can't perform to expectation.

REMEMBER

Although Agentic AI can carry out tasks autonomously, it relies on human design and oversight — as well as interactions with the right technologies — to truly function effectively and responsibly.

Exploring Reasoning, Memory, and Goal Setting

In this section of the chapter, I talk about how Agentic AI systems think, remember, and plan. In essence, Agentic AI systems use three cognitive processes that mimic human intelligence:

>> Reasoning through problems

>> Remembering and applying past and present relevant information

>> Setting and pursuing goals

This mimicking of human thought processes is entirely new territory for machines. Make no mistake, this astounding technological achievement took longer than you may think to arrive. Agentic AI is the natural evolution of decades of work in autonomous and multi-agent systems. Specifically, the branch of AI focused on machines that can perceive their environment, make decisions, and act on their own. The *agentic* label (popularized by social scientist Albert Bandura in the 1980s) borrows from psychology's concept of *agency*, which means the ability to act intentionally rather than just react.

TIP

Oversight of Agentic AI systems requires setting goals and guidelines for agents carefully and supervising the agents vigilantly. The agents don't think, per se, but rather mimic human thinking. Although human thinking is fallible, the processes that pass as machine thinking are sometimes even more so.

Understanding how these machine-thinking processes work can help you grasp both the power and limitations of current AI agents.

Assessing Agentic AI reasoning

Agentic AI reasoning refers to the system's ability to process information, analyze situations, draw conclusions, and determine appropriate actions. This complex computational process can produce remarkably sophisticated problem-solving behavior.

But the AI reasoning capability doesn't include other human abilities that people routinely use when solving a problem. These added and distinctively human superpowers include, but are not limited to

>> **Creative intelligence:** Generate new ideas and solutions to problems by combining originality, flexibility, and problem-solving skills across various domains.

>> **Intuitive intelligence:** Make decisions and solve problems instinctively, relying on gut feelings and subconscious processing.

>> **Moral intelligence:** Distinguish right from wrong and act based on ethical values such as integrity, responsibility, forgiveness, and compassion.

>> **Intrapersonal Intelligence:** Self-awareness and understanding one's own emotions, motivations, and goals.

>> **Naturalistic intelligence:** Recognize, categorize, and draw upon features of the environment.

>> **Emotional intelligence:** Recognize, understand, and manage one's own and others' emotions.

>> **Existential intelligence:** Ponder deep questions about existence, life, and death.

>> **Musical intelligence:** A sensitivity to sounds, rhythms, and music, which musicians, composers, and conductors typically have to produce the *it factor*.

TECHNICAL STUFF

The *it factor* is the elusive spark of emotional authenticity and creative intuition that makes a performance feel alive. By contrast, machine-made music often lacks that essence. It may be technically flawless, but it *lies flat*, meaning the music fails to stir emotion or truly resonate with human hearing.

Considering AI's limited intelligence

The type of intelligence that AI seeks to replicate is *analytical intelligence,* which is the ability to analyze, evaluate, and manipulate information to solve problems and make decisions. However, AI doesn't use the full breadth and scope of analytical intelligence components, which include critical thinking, logical reasoning, abstract thinking, and problem-solving.

REMEMBER

Analytical intelligence doesn't perform at its fullest in a vacuum. When other forms of human intelligence (listed in the preceding section) are also in play — meaning combined and working with analytical intelligence, especially in the problem-solving and decision-making aspects — you get enhanced results derived from analytical intelligence.

Now, you may be thinking "Wait a minute! IBM's AI won against expert players at the game of chess using only analytical thinking!" Yes, indeed — IBM's Deep Blue did that. But that's comparing apples to oranges. Deep Blue was initially built specifically as a chess engine and performed under precise conditions in a highly constrained environment.

Narrow AI successes — such as chess wins — don't provide proof that the AI can replicate the same level of success when solving general or disparate problems in varying conditions and without the benefit of a constrained environment. However, these successes do show that narrowly focused AIs are quite good at what they do. To this day, an AI specialized for a certain task can often outperform a general use model, but only within its intended scope.

Recognizing its intentional design

This idea of task specialization (see the preceding section) clarifies the design of Agentic AI systems that deploy multiple AIs; each agent is specialized to its part of the system's overall mission. Because of this design, many people and companies place more value on employing Agentic AI than on simply evolving GenAI models to higher performance standards.

The people using these technologies are betting that an orchestrated team of specialized AIs can outperform a general model with added capabilities for executing well-defined work processes. This may seem to be a safe bet, but Agentic AI is in its infancy, so the people and companies need to wait and see how well this team approach to autonomous specialized AI plays out before investing too heavily in it.

Evaluating Agentic AI memory

Agentic AI memory refers to its capability to store and recall past data and workflows that it can then use to inform present decisions and future predictions.

People think of Agentic AI memory as working similarly to how a person remembers, which can help them understand the broader concept but doesn't tell the whole story.

REMEMBER

In reality, AI doesn't remember in the human sense; instead, it functions as structured data retention and retrieval. Memory stores information that AI agents can retrieve to build continuity, maintain context, and behave in ways that appear intelligent and responsive over time.

In the absence of memory, every AI interaction would start from the beginning point with no awareness of what happened previously, and therefore, without benefit of the data gleaned from previous actions to draw upon. Consider how much you can enhance system performance just by making memory a capability. For example, agents can remember user preferences, refer back to earlier steps in a task, or adjust their behavior based on the outcome of a prior decision or a change in conditions. This is collectively referred to as the agent *learning over time.*

Adding memory to a system's design

Adding memory to an AI entails more than adding a device that has search functionality. Specialized technologies and tools support the Agentic AI memory by handling the storing, retrieving, and updating of information. (I talk about these types of tools in the section "Identifying Agentic AI building blocks," earlier in this chapter.)

In platforms such as LangChain, memory modules such as ConversationBuffer Memory or VectorStoreRetrieverMemory help manage dialogue history and relevant knowledge. These tools allow an agent to access and reuse earlier conversations, a capability that's especially important for long-form tasks such as writing, customer service, or complex decision-making.

TECHNICAL STUFF

In platforms that have memory modules, the AI model or the AI system often transforms the stored information into *vector embeddings* — which are mathematical representations of words or ideas — that the system can retrieve as related content, even if the prompt or the query wasn't phrased exactly the same way each time. For example, retrieving information about *customer complaints* when your instructions ask about *user feedback* is important even when the exact words don't match.

Agentic AI systems often pair vector databases, such as Pinecone or Facebook AI Similarity Search (FAISS), with memory modules to manage larger volumes of information. Vector databases can help an agent search across documents, chat history, or contextual knowledge quickly and efficiently. For example, an agent might use Pinecone to retrieve previous customer support tickets that are similar to the current inquiry, which can help the agent provide better answers based on past outcomes.

Blending memory and reasoning in the design

More advanced agentic architectures use patterns such as *ReAct,* a term that's short for *Reason+Act,* which blends memory with real-time reasoning. In this approach, the agent remembers previous actions taken and the results those actions produced. The agent can then use that information to make assessments and decisions to guide its next steps. This type of design can really provide a big advantage in flexibility and responsiveness in dynamic environments — where change happens continually — such as when an agentic system troubleshoots software or performs research in which outcomes change based on each interaction.

Without question, including memory in Agentic AI provides substantial benefits in performance. Interactions between AI agents and humans feel more natural and less repetitive. If you ever encountered a system that requests the same information from you several times throughout the process, even though you gave it the same information every time it asked, you probably see this development as great news.

Similar to how Agentic AI can remember what it learned before, it can improve its own process for completing tasks because it doesn't have to repeat steps it has already taken or now understands.

The memory/reasoning blend also helps an agent complete complex tasks in which previous steps influence what happens much later. For example, if an agent is executing a marketing campaign, it must remember which assets are approved for public distribution and which are still in the draft stage.

Getting (too) personal with memory

Agentic AI can remember who you are and distinguish your personal habits, goals, or preferences from those of a zillion other people with whom it may interact. This capability is called *personalization,* and because of memory and better context, Agentic AI can take personalization to a deeper level. Whether that extra insight makes the Agentic AI seem helpful, flattering, or creepy to you or your customers depends on how well the people who made the AI defined its mission and guardrails, as well as your or the end user's perceptions, and not usually on the AI's interpretation of its mission. But there are occasional exceptions.

And that potential creepiness factor brings us to the downsides of Agentic AI memory. Consider these factors:

>> **Systems that use only short-term memory reset after each session —** unless developers explicitly program them to retain information across time. System memory loss can upset or frustrate people who expect the agent to remember things it was never designed to recall.

>> **Long-term memory capabilities can exacerbate privacy and security concerns,** build biases in the AI, or even add confusion in the AI's decision-making processes. Data management, always a critical issue in AI, becomes increasingly important in Agentic AI because the AI memory may not include subsequent policy changes and data updates.

In short, memory transforms an agent from a reactive tool to a more sophisticated and context-aware assistant. Memory bridges the gap between a routine, repetitive automation and sustained, intelligent support.

Grasping Agentic AI goal setting

Agentic AI goal setting is the process that a developer or user uses to give the agent direction. Goals can be simple or complex. Instead of following explicit step-by-step instructions — like GenAI models do when creating content in reaction to your prompts GenAI — Agentic AI agents interpret broader objectives and determine what specific outcomes to pursue on their own.

Agentic AI systems typically organize their goals in hierarchical structures and adjust goal priorities based on changing circumstances, resource constraints, or new information:

>> **Goal structure:** Systems break down high-level strategic objectives into tactical sub-goals and specific actions. For example, a goal to "improve customer satisfaction" might decompose into sub-goals such as "reduce response times" and "increase resolution rates."

>> **Goal priority:** The agent weighs competing objectives and makes trade-offs in real time. For example, a direct shipping route is usually faster, but not if that route cuts through a natural disaster zone or a country where new regulations outlaw the product. In such a case, the agent has to weigh the advantages of a direct route against the emerging challenges on the route and make or change its shipping decision accordingly.

WARNING

Goal setting is a key function that enables AI systems to operate more autonomously in complex, dynamic environments. But the autonomy can raise serious concerns in relation to *business alignment* (adherence to business-wide policies and goals) and keeping the AI agents' actions safe and beneficial for humans. That's why, at least for the foreseeable future, Agentic AI systems must have human oversight and are currently more likely to be semi-autonomous — except for a few restricted use cases (for example, automated stock trading that follows strict rules without human input, factory robots with limited actions, or self-adjusting climate control systems of little consequence).

Understanding Adaptive Behavior and Self-Directed Learning

What sets Agentic AI apart from other forms of AI and AI systems is its adaptive behavior. By *adaptive behavior,* I mean that the AI can self-modify its actions based on changing conditions, updated information, or previous failed attempts to reach a goal. This capability is a distinct departure from rule-based AI systems that follow a fixed sequence of programmed steps, regardless of context, conditions, or outcomes.

Self-directed learning, while still limited in current systems, refers to an agent's ability to improve its performance over time based on experience. This process may involve reinforcement learning, fine-tuning on new data, or simply adjusting preferences through accumulated user feedback. For example, a content-generation agent that frequently receives corrections about tone or audience level can incorporate that feedback into future outputs so that it produces more suitable drafts without a user needing to explicitly retrain it for each project.

Together, the capabilities of adaptive behavior and self-directed learning make Agentic AI more useful than an AI that simply executes instructions. In these systems, the agents are adjusting, optimizing, and sometimes even innovating within the scope of their respective assignments. In short, Agentic AI evolves on the fly and of its own initiative.

Dissecting adaptive behavior

Adaptive behavior enables an AI agent to evaluate progress toward its goal, assess whether its current strategy is working, and pivot to an alternative approach, if needed. This flexibility makes AI infinitely more useful in many real-world applications (such as customer service, logistics, and healthcare diagnostics) where new variables often emerge mid-task.

REMEMBER

Agentic AI adaptive behavior relies on

>> **The AI's capacity to recognize patterns in its successes and failures:** It can then use those pattern insights to refine its approach. An agent constantly compares the current state of the task to the desired outcome. If it detects a mismatch, the agent can adjust its plan or tactics accordingly. It doesn't need to restart the entire process or be reprogrammed manually.

>> **A combination of memory, world modeling, and planning capabilities:** Memory allows the agent to remember prior actions, decisions, and outcomes. World modeling helps it simulate possible future scenarios and choose the best path based on logical reasoning. Planning enables the agent to restructure tasks, reprioritize steps, or incorporate new information into its execution strategy.

Delving into self-directed learning

In Agentic AI systems, self-directed learning takes adaptive behavior even further by enabling AI systems to identify what they need to learn without explicit human guidance. Rather than training on a fixed data set and then deploying unchanged, these systems actively seek out new information and experiences that help them improve their performance. For example, an agent might notice gaps in data when trying to answer certain types of questions, then autonomously seek out relevant academic papers or databases to fill those gaps.

Self-directed learning involves sophisticated mechanisms for identifying learning opportunities: *meta-learning* (learning how to learn) and *agentic reflection (self-assessment of performance).* The process follows a path in which the AI must follow these steps:

1. Recognize when its current knowledge base or capabilities are insufficient to achieve its current goal or complete the assigned task.

2. Determine what additional information or skills can provide the most meaningful improvement in its performance.

3. Figure out how to acquire the needed information, access, or skills — or all three.

This learning process might involve experimenting with different approaches to see what works best, seeking out specific types of data or feedback, or even identifying entirely new domains of knowledge that could enhance its effectiveness.

Learning more than new information

In more experimental Agentic AI systems, agents may autonomously refine work-flows by testing various methods. After monitoring the success of each, the agent selects the most effective strategy for future tasks. For instance, an agent performing market research might test several search strategies, such as querying academic sources, aggregating real-time news feeds, or scanning social media. It then prioritizes the most accurate or relevant method going forward.

In addition to workflow, meta-learning can affect other components of how the Agentic AI systems work. It can

» **Require Agentic AI to develop better goal-setting and planning strategies.** The system learns not just how to achieve specific goals, but how to set more effective goals in the first place. The system can then apply the resulting meta-insights across all future goal-setting activities to improve autonomous planning.

» **Enable Agentic AI to develop better communication and collaboration strategies.** For example, the agent may discover that certain types of explanations work better for technical versus non-technical audiences, or that specific approaches to negotiation are more effective in different cultural contexts. In these and other cases, the AI can apply those discoveries or lessons across the board. In other words, these meta-insights then guide its behavior in future interactions, too.

Examining other aspects of meta-learning

Temporal aspects of meta-learning in Agentic AI — by which I mean that the system learns to balance immediate performance with long-term learning objectives — can really accelerate the agent's overall intelligence and adaptability. The agent can have a sophisticated ability to determine when accepting short-term inefficiency is worthwhile to achieve long-term improvement. A calculation of this nature involves developing strategies for maintaining multiple learning processes. It also requires the agent to simultaneously make quick tactical adjustments for immediate needs while pursuing slower strategic learning to adapt to future capabilities.

Environmental adaptation is another crucial aspect of Agentic AI systems, which continuously monitor their operating context and adjust their behavior accordingly. A financial trading agent, for example, doesn't just follow predetermined algorithms. The AI system also

- **» Adapts to changing market conditions,** regulatory environments, and economic indicators.

- **» Analyzes results** by noticing that patterns which previously predicted stock movements no longer provide a reliable basis for accurate forecasts. The agent then initiates a self-correction process in which it may adjust its models, retrain on recent data, or explore alternative indicators to restore predictive accuracy.

- **» Analyzes further** to diagnose the sources of deteriorating performance and test potential remedies. To correct the issue, the agent may develop new analytical approaches or adjust the weighting of key factors before proceeding to validation.

- **» Validates improvements** through rigorous *backtesting* (which bases changes on historical data) and simulated trading to ensure that incorporating new data patterns genuinely enhances performance rather than merely overfitting the agent to past data. Only after demonstrating consistent accuracy and stability does the agent reintegrate the updated strategy into active trading operations.

Directing Agentic AI

If you read the earlier sections of this chapter, you may wonder how you can instruct Agentic AI agents or systems to do your bidding. Prompting comes to mind, I'm sure. But that's not the answer — at least not entirely.

Prompting an AI is a bit like placing an order at a diner, whereas directing Agentic AI is more like delegating a task to your executive assistant and leaving it up to them to figure out how to get that task done. An Agentic AI agent doesn't just take and fill your order, it accepts the mission that you give it and plans how to achieve that mission, what actions to take, where to go for information, and how to adapt along the way. Occasionally, the agent loops you in to check whether its progress is in alignment with what you meant or to get your permission for a key decision before it acts as a safety check to prevent mishaps.

For your part, you don't have to give step-by-step instructions in a prompt. Instead, you state the *intent*, meaning the end goal, of what you want the AI system to accomplish. Occasionally, you may need to give the AI permission to proceed with a specific recommendation or make a decision if the agent has a question of its own. For safety and compliance reasons, the agent may have its own guardrails that require it to check for your explicit permission before proceeding with an action.

In any case, you don't want to just give Agentic AI your high-level objective and call it a day. The agent needs some context so that it can better understand the mission: what needs to be done, why it matters, and how it fits your or the larger organization's goals. Specifically, you need to provide the agent with

>> **Background information:** Including an explanation of constraints and preferences, and clarification about what resources the agent can or cannot access.

>> **Defined success criteria that go beyond simple task completion:** For example, you can offer your expectations regarding quality standards and strategic alignment with business goals.

Talking it over with Agentic AI

An example of an interaction between you and Agentic AI might look like this:

You: Analyze our customer support logs, identify recurring complaints, and propose solutions.

Agentic AI agent: I'll categorize complaints by frequency, then cross-reference them with product updates to see whether recent changes caused new issues. Should I proceed?

As depicted in this exchange, instead of just spitting out a response like GenAI tools do, the agent might first outline its strategy in its response to you. If you answer the strategy question with *yes*, the agent is off to a running start in getting the work done. If you say *no*, and then explain what you want in more detail, the agent processes that new explanation, provides you with its updated plan, and asks permission to pursue actions on that path.

After you give the agent the go-ahead, it then begins its work autonomously or semi-autonomously, depending on how the agent or system is built to perform. It might query a live database, search the web, retrieve documents, or interact with other software systems, including other AIs, depending on what you empower it to do through your prompting.

REMEMBER

Some agentic systems lean on their memory components to streamline their work on this mission. Others may rely on real-time reasoning to resolve ambiguities or fill in missing pieces. What's most striking is how they don't just follow a script; they improvise. If a source is missing, they look for alternatives. If data is inconsistent, they can flag the issue for you and may even recommend how to resolve it.

Continuing direction over the AI's work

After setting the Agentic AI agent on an agreed-upon path (see the preceding section), you still have work to do. At key decision points, the agent pauses its work and asks for a decision and/or permission to proceed. These intermittent interactions might look like this:

> Agentic AI agent: Sixty percent of complaints are about slow performance, but I'm not sure if this is a server issue or a user interface problem. Should I investigate further?

> You: Yes, investigate further to determine the cause or causes of the issue and provide the sources you used to make that determination,

> or

> No, the cause of the issue is not relevant for my needs.

TIP

Answer interim queries from the AI agent however you want so that the agent knows how to proceed from there. This back-and-forth interaction between you and the agent transforms AI from a passive tool (that responds to queries) into an active collaborator (that assists in solving problems).

Completing the mission and next steps

The level of oversight that you want or need with Agentic AI varies depending on the task, the risks, and your preferences. Some people prefer a hands-on role, requiring the agent to check in frequently, while others are comfortable letting the agent proceed without interruption and only want the agent to notify them when it has the final product ready.

To be clear, the system may respond with summaries, offer previews, or request decisions through dashboards, messaging platforms, or direct interaction in a natural language interface.

After the agent believes that it has reached the mission goal, it delivers the output and often suggests what to do next. A well-designed agent doesn't just drop off a report and vanish; it might follow up with a question like these examples:

>> Would you like me to format this as a presentation?

>> Should I schedule a meeting with the top vendor I identified and prepare a list of items for you to discuss?

Interacting with GenAI and Agentic AI

GenAI and Agentic AI have several key differences in the way that you communicate with them:

>> **Interaction:** GenAI offers a one-shot response, but Agentic AI provides dynamic multiple steps.

>> **Autonomy:** GenAI has low autonomy, simply reacting to prompts, whereas Agentic AI has high autonomy, planning and acting independently.

>> **Feedback:** With GenAI, you have to perform manual retries. On the other hand, Agentic AI uses built-in checkpoints to pause, review its own work, and fix mistakes on the fly.

Table 2-1 offers a concise look at the differences in interactions.

TABLE 2-1   **Interacting with GenAI versus Agentic AI**

Aspect	GenAI	Agentic AI
Method of interaction	Prompting	Instructing
Type of interaction	One-shot response	Multi-step, dynamic
Autonomy	Low (reacts to prompts)	High (plans and acts)
Feedback	Needs manual retries	Built-in checkpoints to monitor progress, evaluate outcomes, and self-correct

Combining Generative Abilities and Real-Time Decision-Making

In this chapter and throughout this book, I compare Agentic AI to GenAI to more easily explain how Agentic AI works and differs from the more familiar and very popular Generative AI models and tools. But in practical use, you don't have to choose only one type of AI. Agentic AI isn't replacing GenAI. It's more like Agentic AI is giving GenAI an upgrade. Think of GenAI as the part that creates ideas and content, and Agentic AI as the part that plans, decides, and follows through. In other words, Agentic AI adds memory, reasoning, and goal-setting so the system can work toward results over time instead of just reacting to a prompt.

Nearly everyone — from students to CEOs — has experimented with GenAI tools such as ChatGPT, Claude, and Grok, experiencing firsthand these tools' remarkable ability to produce coherent, contextually appropriate outputs. These systems excel at pattern recognition and recombination by drawing from vast training datasets to help them generate plausible responses to prompts.

However, this strength also acts as the GenAI systems' fundamental limitation. They operate within the constraints of immediate stimulus and response. They lack any persistent sense of purpose or context beyond the current interaction or a smattering of instruction in its limited memory. Any actions taken on their responses must be manually undertaken or automated through another software program that the AI is integrated with.

If you want to find out more about GenAI models and tools, ChatGPT specifically, or advanced prompting techniques, check out my books *ChatGPT for Dummies* and *Generative AI for Dummies* (Wiley). I also teach several online courses about AI and prompting for business and creative uses at LinkedIn Learning (www. linkedin.com/learning).

Expanding on content generation

Agentic AI builds on a GenAI foundation, adding autonomous action to creative expression. These agentic systems inherit the ability to produce coherent, human-like language and ideas from GenAI, but Agentic AI goes much further by making decisions, setting goals, and executing plans across time and context. This isn't just a technical upgrade, it's a functional evolution in what AI can do.

Specifically, Agentic AI weaves together two critical capabilities:

>> **The creative and generative power of LLMs.** The generative aspect of large language models (LLMs) remains essential. It allows Agentic AI systems to interpret human intent, draft plans, summarize information, produce content, and communicate in nuanced, context-aware ways. Without this generative engine, Agentic AI would lack fluency, flexibility, and the ability to improvise responses or understand the mission it was given. But having these abilities doesn't constitute agency.

>> **The sophisticated real-time decision-making frameworks that enable independent goal pursuit.** The integration of real-time decision-making processes is what transforms an AI system's generative capability into genuine agency. Rather than static rules or pre-scripted routines, these processes are dynamic reasoning loops that provide ongoing evaluations of what's happening (reconsidering assumptions), what might happen next (adjusting strategy), and how best to proceed toward a defined objective (selecting appropriate action).

REMEMBER

This type of reasoning is context-rich. An Agentic AI agent doesn't simply choose the next best word in a sentence in the way that a traditional LLM or GenAI tool does. Instead, it

>> **Chooses the next best action** in a sequence that may span hours, days, or longer. It considers trade-offs.

>> **Balances speed against quality** and short-term outcomes against long-term impacts.

For example, an agent might recognize that providing an immediate answer can solve the current customer service ticket, but if it takes a bit more time to uncover a pattern in the tickets, it can find a common cause to address for preventative purposes or improvement to system-wide performance.

The synergy between these capabilities creates systems that can engage in complex, multi-step problem-solving while maintaining coherent long-term strategies. For example, an Agentic AI tasked with research might generate novel hypotheses through its creative capabilities, then use its decision-making framework to determine which experiments to prioritize, how to allocate resources, and when to pivot based on emerging results. In this scenario, generative power (GenAI) is responsible for ideation, while decision-making (Agentic AI) guides execution.

Applying agentic capabilities to complex interconnections

Agentic AI is new and still evolving, but it can go to work now on more than just theoretical use cases. In enterprise environments, AI teams and technology vendors are already developing agentic systems that can handle workflows that involve dozens of interconnected applications and unpredictable dependencies. For example, an AI assistant tasked with preparing a monthly performance report may need to pull data from different sources, reconcile inconsistencies, interpret anomalies, format the content, and even e-mail the results. And it needs to do all of that while keeping the report's owner in the loop and adjusting its plan if a dashboard goes offline or a new dataset becomes available.

REMEMBER

In this use case, the agent needs at least a degree of autonomy so that it doesn't just wait to be told what to do next. It notices, adapts, and keeps working toward the goal of creating a timely and accurate performance report.

Operating autonomously across time

What makes the evolutionary step for AI so transformative is that Agentic AI can operate across time:

>> **Traditional generative models excel in single-session contexts.** They can generate impressive content, answer questions, and even simulate dialogue. But they don't persist. They don't remember what they did yesterday, and they don't track what needs to happen tomorrow.

>> **Agentic systems, by contrast, maintain continuity.** They track tasks, store knowledge across sessions, and use memory to reflect on what worked, what didn't, and what they still need to do. They're not just intelligent, they're persistent.

The agentic systems' combination of intelligence and persistence enables sophisticated tool use and interaction with external systems. Agentic AI can generate appropriate commands, API calls, or interface interactions while simultaneously monitoring the results and adjusting its approach. In doing so, it functions more like an orchestrator than an assistant by

>> Not simply responding to requests, but coordinating a series of actions that often run across various software platforms

>> Working to achieve an outcome that's clearly defined but not tightly scripted

This design flexibility makes Agentic AI particularly well-suited to environments in which uncertainty is the norm rather than the exception. Business, science, healthcare, and logistics are all domains filled with ambiguity and change. Success in these areas depends not on executing a single correct answer, but on navigating evolving conditions and revising plans when new information becomes available.

Staying the course in a changing environment

The marriage of content generation and ongoing decision-making enables Agentic AI systems to operate with genuine autonomy in complex, unpredictable environments.

Nothing (and no one) can successfully create and make decisions in a linear fashion. Agentic AI's nonlinear approach isn't a flaw; it's a feature. It reflects the messy, adaptive nature of real-world problem-solving. In many ways, the process mirrors how human beings work when they're pursuing open-ended goals. They gather information, consider options, make a move, and reassess. They change course when needed, seek help when appropriate, and keep moving toward the finish line. Agentic AI follows the same process, only faster, more consistently, and at greater scale.

REMEMBER

Agentic AI uses the blend of GenAI and a solid decisioning framework to engage with the real world. It uses language and logic as tightly integrated tools in the service of action. That integration, the fusion of creativity and control, improvisation and intention, makes these systems so powerful, and so promising for real-world problem-solving and innovation. Table 2-2 offers a comparison of GenAI and Agentic AI capabilities.

TABLE 2-2

Comparing GenAI and Agentic AI

Feature/Aspect	GenAI	Agentic AI
Language/creativity	Yes	Yes (inherited and enhanced)
Decision-making	Limited/static	Dynamic, real-time, adaptive
Goal pursuit	No	Yes (sets and pursues goals)
Execution	No	Yes (executes plans over time)
Adaptability	Limited	High (continuously learns and adjusts)

Generated with AI using Perplexity AI

IN THIS CHAPTER

» Focusing on coordination
and planning

» Developing contextual awareness
and situational reasoning

» Processing self-directed
improvements

» Adapting for multiple inputs
and systems

» Considering common protocols

» Choosing options in agent
construction

Chapter **3**

Meeting Agentic AI Core Technologies

The rise of Agentic AI marks a major turning point in the evolution of artificial intelligence (AI). This level of development represents a dramatic jump for machines, from being simply reactive tools to active participants in decision-making and goal execution. (Chapter 2 offers a big-picture look at this evolution.) But what powers this leap from prompt to pursuit? What technologies make true agency in AI systems possible?

This chapter unpacks the core technologies that enable Agentic AI to function with purpose and persistence. From world models and memory systems to real-time reasoning loops and tool orchestration, I help you explore the underlying frameworks that give these systems their intelligence, adaptability, and operational independence. This information offers you a peek at what's under the Agentic AI hood and why it matters. But remember that new programming developments can change this technology fast — and agents can make changes themselves, too.

Driving Multi-Agent Coordination and Planning

The idea behind multi-agent systems (such as Agentic AI) sprang from the recognition that a single AI agent can't efficiently handle many complex tasks — such as managing supply chains, running simulations, designing and executing scientific research, or orchestrating autonomous vehicle fleets. Instead, complex tasks require a dynamic, decentralized group of agents that can communicate, coordinate, and act both independently and collaboratively toward shared or complementary goals.

This description of a grouping of coordinated agents depicts the idea of Agentic AI systems. But turning the idea into a functioning reality requires quite a technological feat.

Meeting the practical demands of any complex task requires a convergence of several technological advancements that can work together effectively, often within seconds (or less). To perform the complex task, Agentic AI needs a steady, predictable system because even small glitches within it can cause big failures. For example, a warehouse robot that misreads inventory data may send hundreds of boxes to the wrong location, or a financial trading agent may keep buying shares of a stock — instead of selling them — because of one missed signal somewhere along the way.

REMEMBER

All of these directives — meeting task demands, doing so quickly, and maintaining a consistent level of performance — apply to multi-agent coordination and planning in Agentic AI. But also, several technical and practical considerations drive how developers and organizations design, form, and deploy multi-agent systems. The following sections address those considerations.

Computational complexity and task decomposition

Your brain decomposes large goals and projects into smaller tasks or single steps. This segmentation is called *task decomposition,* which breaks up a larger problem into smaller, more manageable tasks. Think of this saying: "What's the best way to eat an elephant? One bite at a time." You can't devour the elephant (or complete the task) in a single bite. This process of decomposing a larger problem into smaller, more manageable tasks is reflected in multi-agent systems for the same reason.

Complex real-world problems typically exceed the computational and functional limits of single AI agents. Multi-agent systems can better address complexity by decomposing large problems into smaller, manageable subtasks that they can tackle in parallel processes to achieve the larger goal.

In other words, the systems take a distributed approach to handling scenarios that would greatly exceed the capabilities of a single AI agent. Examples of such scenarios include coordinating supply chains, managing smart city infrastructure, or orchestrating autonomous vehicle fleets.

Specialized expertise and division of labor

Agentic AI systems mirror how people think and work. (Just remember that mirrors are reflections and not exact duplications of the things that they reflect.) AI agents offer shallow reflections, not complete clones of human thinking.

In the same way that humans create their working groups with generalists and specialists, so do Agentic AI systems. Specialized agents are *enhanced* (optimized) for better performance in specific domains or for specific capabilities. Rather than creating one generalist agent that performs adequately across all tasks, multi-agent systems use specialized agents that excel in narrow domains so that together they perform at much higher levels than the generalized AI agent can do.

For example, one specialized agent might handle natural language processing (for communication), another computer vision (for interpreting the visual data), and a third logical reasoning (for making decisions based on available data and situational awareness). This specialization mirrors how human teams organize around members with different but complementary skills so that the team members together can complete a complex task that is too difficult for one person.

Scalability and fault tolerance

Multi-agent architecture provides *scalability* (the capacity to expand to accommodate additional data or tasks without reducing performance) by adding or removing agents based on workload demands. The architecture also offers strength and resilience through redundancy. If one agent fails, others can potentially compensate or take over its responsibilities. This is particularly crucial for mission-critical applications such as autonomous vehicle fleet management systems or financial trading platforms.

But the scalability advantage goes beyond simple *horizontal scaling* (distributing the work across computing resources). Multi-agent systems can also

» **Implement sophisticated load-balancing tactics.** Agents can dynamically redistribute tasks themselves based on real-time performance metrics.

For example, agents can monitor their own processing capacity and spawn additional worker agents or request task reassignment when demands upon it exceed its capacity. Therefore, multi-agent systems have some elasticity; they can automatically adapt to varying computational demands.

» **Adapt for future processing demands and performance goals.** Agentic AI systems can spawn and retire agents by applying intelligent policies that consider not just current load, but predicted future demands, resource costs, and system performance targets.

By using both internal agent monitoring and intelligent policies, a cloud-based multi-agent system could scale up agents before predicted traffic spikes and then scale down when traffic diminishes — to optimize its use of resources.

Fault tolerance in multi-agent systems operates at multiple levels:

» **At the basic level,** *agent redundancy* means that the system distributes critical functions across multiple agents, so no single point of failure can compromise the entire system.

» **At a higher level,** *hierarchical fault tolerance* involves supervisor agents that monitor subordinate agents. Supervisor agents can reassign tasks, restart failed agents, or reconfigure the agent network in response to agent failures.

Supervisor agents provide redundancy through multiple layers of fault detection and recovery, making Agentic AI systems more reliable than they would be without them.

Distributed information and resources

Imagine trying to manage a global shipping company from a single office in New York. Every decision about a ship in Singapore or truck in Berlin would require information to travel halfway around the world and back, potentially creating long delays and high communication costs. Multi-agent systems eliminate this bottleneck by distributing intelligence and authority directly to local nodes. This distribution allows decision-making to occur where the information resides, rather than waiting for instructions from a distant central command.

When resources can't be moved

Many real-world scenarios involve distributed data sources, sensors, or computational resources that can't move to a centralized location efficiently. Just imagine what could potentially happen if an autonomous car had to send all the data pertaining to immediate driving conditions to a central data center for processing and then get instructions back in time to avoid an impending crash.

While autonomous vehicles run on a different system than Agentic AI, the urgency in matching data to actions is the same. Agentic AI systems can solve this urgency problem by operating in the same location in which the data source or sensor network generates and stores information. In other words, data is generated, processed, and possibly stored at the same location (or nearby). This localized processing reduces communication overhead and latency while maintaining local autonomy for time-sensitive decisions.

Autonomous cars already use localized processing. They achieve this distributed computing through a mix of *edge computing* (right on the vehicle itself or nearby systems like smart roads and smart city networks) and cloud computing.

>> **Edge computing:** Local sensors and edge computing devices on the vehicle can generate and process the immediate machine decisioning. The edge computing part handles urgent, split-second decisions, like detecting an obstacle and braking to avoid a crash.

>> **Cloud computing:** Internet-based; used for non-urgent computations, such as maintenance, vehicle, and driver monitoring.

Apply the multi-agent approach

Agentic AI follows the same approach as autonomous vehicles in processing: a blend of edge and cloud computing. (See the preceding section for more on these types of computing.) But multi-agent systems can provide additional advantages over the current system that autonomous vehicles use.

Predicted advantages for example scenarios incorporate concepts of how AI agents can demonstrate

>> **Broader autonomy and increased adaptability:** They apply learned patterns to varying scenarios and can adjust strategies based on real-time conditions within their operational domain.

>> **Goal-directed behavior:** They proactively plan complex multi-step tasks to meet safety, efficiency, and navigation goals simultaneously.

>> **Increased contextual awareness:** They reason through uncertainty, predict and respond to complex road conditions, and engage in interactions with other agents or smart infrastructure.

>> **Automated scalability and collaboration:** They enable sophisticated vehicle-to-everything (V2X) interactions for cooperative driving, hazard sharing, and ecosystem-level coordination.

Emerging coordination mechanisms

Recent advances in coordination protocols help make multi-agent systems more practical. These advances include

>> **Consensus algorithms:** Methods that help distributed systems (like agents) to agree on what's true or what action to take without the need of a human boss or a central control system to make the call. In simple terms, the algorithms are how agentic systems can vote on a decision and way forward. No one person (a human) or single program (a supervising agent) is officially in charge.

This protocol forms the backbone of reliable coordination in decentralized systems. If you want to know more about how a logical outcome can come from a mix of inputs, consider the Byzantine Fault Tolerance (https://www.geeksforgeeks.org/system-design/byzantine-fault-tolerance-in-distributed-system/) and Raft consensus (https://raft.github.io/) protocols. Both are examples of algorithms that enable agents to maintain consistent worldviews even when some agents fail or provide conflicting information.

- *Byzantine Fault Tolerance (BFT)* is a way for systems to keep working correctly even when some parts fail unpredictably or act maliciously. It's kind of like a group reaching the right decision even if a few members are confused, lying, or actively trying to sabotage.

- *Raft* is another consensus method. It keeps all agents or computers in agreement by electing a leader to coordinate updates. This works like choosing a team captain to make sure everyone's expectations stay in sync. Unlike BFT, Raft assumes that parts might crash or disconnect, but that they won't actively deceive others.

>> **Auction-based task allocation mechanisms:** These provide the means to dynamically distribute work among agents through a competitive bidding process. Having AI agents bid on task assignments may sound bizarre, but this mechanism really offers a practical means of assigning values or weights to different ways of accomplishing the task at hand.

For example, in protocols such as the Contract Net Protocol (https://www.sciencedirect.com/topics/computer-science/contract-net-protocol), agents announce available tasks, receive bids from capable agents, and award contracts based on factors such as cost, capability, and availability. This market-based approach is designed to optimize resource utilization and adapt to changing conditions.

>> **Hierarchical planning systems:** These organize agents into multi-level coordination structures in which higher-level agents direct lower-level specialists. Supervisory agents decompose complex problems into manageable subtasks for specialist agents, assign the subtasks to the right agents, and create clear command chains. These systems leverage local agent autonomy (agents work independently) while also keeping agents on track to achieve coherent global objectives.

>> **Game-theoretic approaches:** Help agents negotiate and collaborate even when they have competing objectives. These protocols provide mathematical frameworks for strategic interaction between agents that have potentially conflicting goals, thus helping agents find stable cooperation strategies, negotiate fair resource sharing, and reach equilibrium solutions, even in competitive environments. Techniques such as mechanism design and cooperative game theory help systems find the sweet spot between what's best for each agent and what's best for the group as a whole.

For example, imagine a fleet of delivery drones all trying to use the same limited airspace. If each drone acted purely in its own self-interest, they'd collide or waste battery life avoiding each other. Game-theoretic rules help them negotiate flight paths and charging times so that each completes its deliveries efficiently and with no midair collision chaos.

TIP

These coordination mechanisms and protocols tend to work well across diverse applications, including autonomous vehicle coordination, distributed computing clusters, supply chain management, and robotic swarm operations.

Communication and shared understanding

Multiple AI agents need to speak the same language and understand each other's messages so that they can work together effectively. Recent advances in related technology — such as natural language processing (NLP); large language models (LLMs); emerging protocols such as Model Context Protocol (MCP), Agent Communication Protocol (ACP), and Agent2Agent (A2A) protocol; adaptive communication networks; and frameworks such as AutoGen — make this communication much more sophisticated and reliable.

Researchers have developed standardized communication protocols that

>> **Allow different AI agents to share information clearly, even if different teams or companies built them.** Developing these protocols is like creating a universal translator that ensures agents can exchange data about what they're sensing, detecting, or observing, what tasks they're working on, and what decisions they're making.

>> **Help agents maintain a shared understanding of their environment.** Examples include a warehouse automation system in which all agents agree on where packages are located, which robots handle which tasks, and what the current priorities are for fulfilling orders or replenishing stock. This shared understanding prevents confusion and conflicts that could arise when agents have different or outdated information.

>> **Allow agents (in newer systems) to communicate by using more natural, human-readable language (known as natural language processing, or NLP) rather than just rigid computer code.** This feature helps humans understand what agents are discussing and decide when to step in (if needed). It also makes agent-to-agent communication more flexible and adaptable to unexpected situations, for example, when a new variable or problem appears that wasn't part of the agents' original programming.

Connecting Contextual Awareness and Situational Reasoning

The interplay between *contextual awareness* (what's happening, where, and with whom or what) and *situational reasoning* (what I should do about it now, and why) elevates multi-agent systems from tool-like assistants to autonomous decision-makers. The core technologies involved combine perception, memory, logic, and coordination to create the foundation for collective intelligence in agentic systems.

The following sections offer you a breakdown of the core technologies that enable the connection between contextual awareness and situational reasoning.

World modeling

World models provide AI agents with a shared representation of their environment. This representation includes spatial, temporal, and causal relationships. Also, world models exist on a spectrum from simple state representations (like a task list, map grid, or game board showing object positions and states, for example) to complex physics simulations (such as virtual 3D environments used for robotics training or autonomous driving simulations), depending on the application requirements and any existing computational constraints. The idea of a shared representation can look different across Agentic AI systems. In some designs, all agents access a single shared world model. In others, each agent maintains its own internal model of the world, which synchronizes or updates periodically through communication with other agents.

AI system architects often construct world models by using large-scale simulations, 3D environments, or abstracted state representations. Constructing world models can be quite expensive and resource-intensive, although the investment varies significantly based on the architects' approach. The most expensive world models to construct are large-scale simulations and 3D environments. The least expensive to construct are abstract state representations (but creating those world models isn't cheap and still takes time and requires extensive testing).

Developers and architects invest in these complex world models because they allow agents to learn, plan, and test decisions safely before acting in the real world, which ultimately reduces costly mistakes and improves performance. See Chapter 10 for more information about world models.

In short, world models allow agents to

>> Anchor their understanding of context in specifics such as the layout of a building, the sequence of events, the laws of physics, and object locations

>> Predict consequences of their own and others' actions, whether those others are agents or people

>> Plan and reason over time, not just space

For example, in disaster response or in simulations, multiple drone agents can rely on a shared or synchronized world model to navigate debris-filled environments, anticipate collapses, and coordinate with human rescuers.

TIP

Real-world simulations may look impressive, but they aren't always the best choice for educating agentic systems. The more detailed a world model is, the more computer power and time you need to run it — sometimes without adding much real benefit. Start with simpler models when possible and increase complexity only when the need for accuracy or realism increases. A lightweight, well-tuned world model often performs better (and faster) than an overly detailed one that bogs down your system.

Perception and sensor fusion

Agents must perceive their environment through diverse inputs such as cameras, microphones, *light detection and ranging* (LIDAR, which uses pulsed laser light to measure distances), IMUs (inertial measurement units), GPS, infrared sensors, tactile sensors, as well as digital sources like text feeds, APIs, and network data in order to be contextually aware. Sensor fusion merges these inputs into a coherent and usable understanding of the environment.

This merged perception of the environment enables agents to

>> Adapt to environmental changes in real-time.

>> Recognize specific meanings or objects by the association of symbols, such as recognizing that a phrase such as *the blue box* refers to a specific visual object.

>> Detect emergent patterns and anomalies that may elicit a corrective response from the agent.

For example, in smart manufacturing, robotic agents fuse sensor data from vibration, heat, and visual inspection to detect equipment failure and alert the appropriate people before downtime occurs.

REMEMBER

Sensors to include as input depend on the context of your system, so for

>> **Autonomous vehicles,** use LIDAR, radar, cameras, IMU, and GPS.

>> **Industrial robots,** add tactile, force, and vibration sensors.

>> **Software agents,** focus on APIs, databases, and network data.

>> **Drones,** include IMU, barometric pressure, and GPS.

Memory architectures

Multi-agent systems need memory to track context over time and across tasks, conversations, or missions. This includes three types of memory:

>> **Episodic memory:** To recall specific events or experiences

>> **Semantic memory:** To store general knowledge or learned facts

>> **Shared memory:** To allow collaboration and information exchange between agents

Memory architectures support situational reasoning, such as remembering that Agent A failed a task yesterday (episodic memory) due to low battery (semantic memory) which should influence today's plan (shared memory) to include that the battery be recharged or replaced.

Theory of mind modeling

To coordinate with other agents or humans, each agent in an Agentic AI system must maintain a model of the others' beliefs, goals, and likely actions. This modeling is sometimes called *recursive reasoning* or *theory of mind*.

This capability enables agents to

>> **Anticipate partner or adversary behavior.** For example, when one car in a fleet of autonomous vehicles slows down to let another car merge. No person (nor thing) told the car that slows down to do so, but it inferred the other car's intent and acted accordingly.

>> **Avoid redundant work.** For example, by circumventing certain tasks that they expect another agent to perform. In a multi-agent maintenance system, one agent may detect that another is already diagnosing a malfunction and redirect its own effort to a different subsystem instead.

>> **Strategize in cooperative or competitive settings.** For example, in resource allocation, negotiation, or multi-agent games, each agent may predict how others will act and adjust its strategy to maximize overall success or achieve an advantage.

REMEMBER

The *theory of mind* in AI refers to the capacity of an AI agent to understand that other agents have characteristics that may differ from their own. Keep in mind that agents don't actually understand thoughts or intentions the way humans do. Instead, they use patterns and probabilities to approximate what another agent or person might do next. These models don't create real empathy or awareness, but instead, use clever calculations that make coordination smoother and decisions smarter.

Communication protocols and intent signaling

Multi-agent AI systems require sturdy and reliable inter-agent communication frameworks, which are often augmented by natural language, application programming interfaces (APIs), or symbolic signals. This type of framework includes

>> **Negotiation protocols** that enable agents to propose, bargain, and agree on plans or resource allocations so that they can resolve conflicts or balance competing goals without human intervention.

>> **Task allocation mechanisms** in which agents decide which agent should handle which task based on their capabilities, availability, or current workload. This prevents duplication of effort and optimizes efficiency.

>> **Status updates and goal sharing** messages that agents use to announce progress, signal intent, or synchronize their actions with others, thereby helping the system adapt dynamically as conditions change.

In the context of multi-agent agentic systems, communication doesn't only involve passing information, but it also makes that information actionable in the current situation.

Planning and goal-conditioned learning

Agentic AI implementation for systems such as robotic fleets, autonomous drones, or adaptive customer-service agents, often need to adapt agent behavior when goals change or unexpected events occur. In these cases, agents use hierarchical planning or goal-conditioned reinforcement learning to take context-aware actions based on goals and current situations.

This planning and learning includes

>> **Decomposing tasks into subgoals** when a problem is too large or complex for a single agent to solve directly. For example, a delivery drone might break down a long route into smaller waypoints to handle navigation more efficiently.

>> **Adjusting plans** when environments or other agents behave differently than expected, prompting supervising agents to reassign priorities or modify strategies so the team can still reach the overall goal.

>> **Reasoning through what-if scenarios in real-time** for the purpose of anticipating outcomes, preventing errors, and choosing the most effective next action. For instance, an autonomous drone might simulate different flight paths in milliseconds to find one that avoids bad weather and conserves battery power.

Distributed coordination and federated learning

In large systems, individual agents may train locally while learning from shared experiences. Federated learning and distributed coordination ensure consistency without centralization which is important in time-sensitive or decentralized environments such as healthcare networks, smart city infrastructures, or autonomous vehicle fleets.

This learning and coordination supports

>> **Scalability** by allowing many agents to train and operate in parallel without overloading a central server or communication channel.

>> **Adaptability** to local contexts for fine-tuning behavior based on regional data, user needs, or environmental conditions while still aligning with the system's global objectives.

>> **Continual learning** across agents and teams, which keeps the Agentic AI systems in synch with the current operating environment and able to evolve as new data, rules, or goals emerge.

Self-Correcting Continuous Improvement

In Chapter 1 of this book, I note that multi-agent Agentic AI systems are collections of autonomous AI agents that work together, learn from experience, and adapt while they go. Unlike traditional software that runs on fixed instructions, xxx AI developers design Agentic AI systems to learn and adjust their behavior in real-time.

At the heart of this learn-and-adjust improvement process, you find feedback loops. Just like you might rethink a strategy after a failed attempt at completing a project, agentic systems monitor their own actions and results to determine what they did right and what might have gone wrong. When something doesn't go as planned, they analyze what happened, adjust their reasoning, and try a new approach.

Improving by failing and adjusting

Suppose that an agentic AI system is tasked with keeping deliveries on schedule when a sudden storm hits. One agent might try rerouting around the weather, only to discover that the detour causes even longer delays. That failure isn't wasted — it's a learning opportunity. The agent remembers that this particular detour doesn't work and updates its internal model. It then shares that update with other agents so the entire system can avoid repeating the same mistake.

This ability to share insights and lessons learned gives multi-agent systems several key advantages:

>> **Collective intelligence:** Learned lessons spread across the network instead of staying trapped within one agent; this sharing creates a smarter overall system.

>> **Decentralized coordination:** The agentic system doesn't need a central brain to distribute new knowledge. Agents coordinate the transfer of information through shared memory (see "Memory Architectures" earlier in this chapter), messaging protocols (see "Communication and Shared Understanding"), and sometimes even a shared world model that helps them understand the big picture (see the section "World modeling," earlier in this chapter).

>> **Adaptive reasoning:** Each agent can reason based on this shared context, predict outcomes, and fine-tune its actions accordingly.

TECHNICAL STUFF

HOW AGENTS SHARE THEIR LESSONS

When an agent learns from failure, it's not just taking mental notes. It's updating data structures or models. Agents can share those lessons in several ways:

- **Policy updates:** An agent adjusts the mathematical rules (its policy) that guide future actions and then synchronizes that policy with peers or a shared repository.

- **Embedding sharing:** Some systems exchange compact vector representations (embeddings) that capture what was learned about a task or environment.

- **Memory synchronization:** Agents periodically update a shared memory store so that others can read new facts, outcomes, or warnings.

- **Federated learning:** Each agent trains locally on its experiences, and the results are aggregated to improve the group's global model (without moving around all the raw data).

Together, these methods give multi-agent systems a kind of hive mind that continuously refines itself through feedback and shared experience.

Correcting more than just mistakes

Agent and system improvement involves more than correcting mistakes (which you can read about in the preceding section). Agentic AI systems also engage in meta-reasoning, which means they think about how they think to refine their decision-making strategies over time, update goals based on new data, and sometimes even reassign roles among themselves to be more effective.

TIP

The ability of systems to improve and adapt autonomously is especially useful in fast-changing environments such as cybersecurity, finance, or emergency response, where static systems quickly fall behind. Multi-agent Agentic AI can thrive in the midst of real-time unknowns, growing smarter and more capable when new challenges arise.

It all sounds a little spooky, I know. But keep in mind that this new technology will improve over time — and that's likely a good thing in the same way that our knowledge of medicine improves over time. The issues of ethics, biases, criminal use, and human responsibilities as they relate to AI systems will remain, of course. These aspects always arise with the use of any new tool (for example, with privacy concerns related to the adoption of the internet, or misinformation risks that arose with social media). You can dive into the discussion of those issues by reading Chapter 7.

Shifting to Multimodal Input and Cross-Domain Functionality

Until recently, interacting with AI meant typing prompts. At least, that activity was the experience of the general public while they hurried to understand how to use Generative AI (GenAI) chatbots after ChatGPT's debut in the mainstream. For many people, GenAI was the only form of AI that they knew, and prompting in human language was the way to interact with it.

Prompts are carefully worded instructions intended to coax a response from a model trained to ingest language inputs to generate outputs. But Agentic AI systems leave that narrow paradigm behind. Instead of waiting for a prompt before they act, they can perceive, reason, and autonomously act across multiple domains by using a wide array of data sources far beyond words.

Contrasting reactive and proactive operation

REMEMBER

The shift from generating simple responses to taking meaningful actions is a hallmark in multi-agent Agentic AI systems, and an astounding feat considering groups of autonomous agents must cooperate to solve complex tasks. These agents don't just take a prompt and respond. They take in information from many inputs, including visuals, audio, structured data, sensors, application programming interfaces (APIs), logs, and yes, even human language. Then they reason about how to achieve their assigned goals in ever-changing environments.

Traditional prompt-based systems are reactive by design, meaning they wait for you to say something before they go to work to respond to that particular stimulus. Agentic systems, by contrast, are goal-driven and proactive. A prompt might kick off a task, but agents continuously seek out new information on their own to refine their strategies and coordinate with other agents, systems, devices, and people in order to undertake meaningful actions, including writing prompts for other agents.

For example, in a multi-agent system that manages an industrial facility, one agent may monitor temperature sensors, another inspects equipment visually via computer vision, and a third pulls data from maintenance logs. Together, they decide whether a machine needs servicing. The system doesn't require a prompt; no one had to ask. The agents infer and reason their way through to a conclusion and then act. Examples of actions the agents may take include scheduling one or more service technicians, ordering replacement parts, or noting in a log that all is well and no servicing is needed at this time.

Exploiting multiple streams of input

The shift toward multimodal input means Agentic AI systems can perceive the world in a way that's more human-like. For example, they can

>> **See** through cameras and visual data

>> **Hear** through microphones and audio streams

>> **Read** structured and unstructured data

>> **Sense** environmental variables from Internet of Things (IoT) sensors

>> **Interpret** spoken or written human instructions when needed

Multimodal agents integrate diverse inputs into shared mental models, grounding abstract goals in concrete reality. This fusion of perception and reasoning helps agents form context-rich decisions, adapt to ambiguity, and act autonomously in unfamiliar or unpredictable scenarios.

Multi-agent systems also perform as cross-domain actors by using this well-informed foundation. *Cross-domain* means that a single network of agents can operate across multiple industries, functions, or knowledge domains without being hardcoded for each one.

In healthcare, for example, one agent might monitor vitals in ICU rooms, while another tracks and monitors patient equipment and medical supply usage, and a third pulls patient data from electronic records to detect anomalies or risks. Together, these agents form a cross-functional AI team, capable of reasoning across silos that humans often struggle to bridge.

Enabling the growth of multimodal systems

Shared knowledge graphs, flexible APIs, common data standards, and world models that abstract knowledge form the backbone of evolving multimodal systems. As agents become better at translating insights from one domain to another, they start to display *generalizable intelligence,* meaning skills that apply beyond a single task or context. This kind of adaptability is often described as a step toward artificial general intelligence (AGI), the kind of thinking machines long imagined in science fiction genre, such as *Ex Machina, Her,* or even Data from *Star Trek: The Next Generation.*

But science fiction aside, you and I — and the rest of the world — are only beginning to glimpse what's possible with Agentic AI systems and whatever follows. Even so, the trajectory is clear: the future of AI interaction won't just involve humans typing prompts and machines spitting out text. Instead, both sides will

>> Sense, reason, and respond to one another within shared contexts.

>> Understand tone, environment, intent, and even emotions in others.

>> Share context that emerges from a network of intelligent agents working together across disciplines, systems, and everyday realities.

Does that mean that people are close to bringing science fiction-level AGI to life? No, it doesn't. That kind of superintelligence requires more advanced technology than humans have now. Agentic AI is just a step, albeit an important one, along that path.

SCI-FI AGI VERSUS AGENTIC AI

TECHNICAL STUFF

Science fiction genres have been dreaming up artificial general intelligence (AGI) for decades. And while the movie versions are often exaggerated for drama, they've done a surprisingly good job of exploring the what-ifs that real researchers now face.

- **Ex Machina (2015)** — A sleek, unsettling look at consciousness and manipulation. Ava's ability to plan, deceive, and adapt makes her one of the most realistic depictions of an agentic system gone rogue, though today's AI isn't even close to that level of self-awareness.

- **Her (2013)** — A gentler take on conscious AI. Samantha learns, reasons, and forms emotional connections. This is more or less what future agentic systems may do, minus the heartbreak.

- **Data from Star Trek: The Next Generation** — The fan-favorite android who strives to understand humanity. Data captures the spirit of generalizable intelligence: learning and applying knowledge across domains, not just completing a single task.

- **I, Robot (2004)** — A flashy action movie loosely based on Isaac Asimov's classic stories about ethics and autonomy. It reminds us that intelligence alone isn't the issue; alignment with human values and full accountability are.

While none of these movie portrayals are fully realistic, they raise the right questions: What happens when machines not only follow commands but make their own judgments? That's the frontier Agentic AI is beginning to explore carefully, and (hopefully) with more humanity than Hollywood usually allows.

Streamlining Integrations Using New Protocols

While organizations deploy multiple AI agents across various platforms and use cases, AI developers increasingly see the need for a mature AI ecosystem that has unified communication protocols. The AI landscape is approaching a pivotal point as the industry works to move away from ad hoc integrations and toward interoperability and standardized protocols for AI agent communication. After years of fragmented, proprietary solutions, four major protocols have emerged as frontrunners in the race to standardize how AI agents interact with tools, data, and each other.

Currently, the industry is leaning toward lightweight protocols such as Model Context Protocol (MCP), Agent Network Protocol (ANP), Agent-to-Agent (A2A) protocol, and Agent Communication Protocol (ACP) which you find out more about in the following sections. For now, suffice it to say that these new protocols represent a significant shift from earlier proprietary approaches.

The protocols address critical challenges in AI deployment which include

>> **Connecting powerful language models to diverse data sources.** Large language models (LLMs) are powerful reasoning engines, but on their own they lack direct access to real-time or domain-specific information. Protocols such as the Model Context Protocol (MCP) bridge that gap by standardizing how LLMs securely connect to external tools, APIs, and databases. This enables agents to ground their reasoning in up-to-date facts, enterprise data, and contextual knowledge without hard-coding every integration.

>> **Enabling secure multi-agent collaboration.** As agentic systems evolve, multiple autonomous agents often need to cooperate across different organizations or domains. The Agent Network Protocol (ANP) and related frameworks address this by providing identity, authentication, and trust mechanisms (often using decentralized identifiers, or DIDs). This ensures that agents can discover one another, share tasks or data, and coordinate actions while maintaining verifiable provenance and privacy boundaries.

>> **Creating scalable integration patterns.** Traditional APIs and ad hoc connectors don't scale well when thousands of agents, models, and data systems need to interoperate dynamically. These emerging protocols introduce standardized, reusable integration layers that support plug-and-play extensibility — so developers can add new agents, capabilities, or data sources without rebuilding entire pipelines. This scalability is essential for the next phase of AI deployment, where networks of agents act collaboratively across diverse infrastructures and contexts.

Model Context Protocol (MCP)

The *Model Context Protocol,* or MCP, is an open-source protocol designed to solve a fundamental challenge in artificial intelligence: how to securely and efficiently connect AI models to external systems and data sources. At the time of this writing, wide variations in implementations exist, and in most of them, security receives little more than an afterthought. Even so, MCP is in use now and is currently the front runner in the standardization race.

Created by AI safety and research company Anthropic and open-sourced in late 2024, MCP emerged as a response to the growing need for a standard way for

AI-powered tools to communicate with external data sources, tools, and services. While AI systems take on more dynamic roles in enterprise and developer environments, the lack of a consistent way to supply real-time context to large language models (LLMs) becomes a notable bottleneck. MCP aims to fix that.

REMEMBER

MCP functions like a universal adapter — sort of like a USB-C port for AI. Instead of developers building custom application programming interfaces (APIs) or one-off data connectors for every tool or model, MCP offers a single, shared interface that any application can use to provide context to an LLM. This access allows AI applications to request data, trigger tool usage, and share situational context in a streamlined format.

Seeing how MCP works

MCP works by using a common setup found in many digital systems: a client-server model. Think of the setup like a conversation. One side (the client) asks for something, and the other side (the server) provides it. In this case, the AI agent or tool is the client, and the data source or external service is the server. They communicate by using a lightweight digital language called JavaScript Object Notation Remote Procedure Call (JSON-RPC), which helps them clearly understand each other's requests and responses. This communication setup

>> **Gives developers a reliable and consistent way for AI tools to request data or trigger actions without having to build a custom connection every time:** The interaction isn't always as simple as plugging in a USB drive, but compared to traditional integrations, MCP greatly reduces the complexity and makes feeding up-to-date, useful information into an AI system much easier.

>> **Demonstrates growing industry support:** Microsoft, for instance, has added MCP capabilities to its Copilot Studio and is integrating it into its Azure AI Agent Service via Azure AI Foundry. This kind of enterprise adoption signals a high degree of confidence in MCP's long-term utility and durability, particularly for developers building at scale.

Recognizing MCP's limitations

MCP does have its limitations. Despite being the most mature among emerging AI communication protocols, its current scope is relatively narrow, as indicated by its

>> **Lack of backing for complex interactions:** MCP excels at connecting models to tools, but it doesn't support more complex peer-to-peer or agent-to-agent interactions. Currently, A2A — which you can read about in the later section

"Agent2Agent protocol (A2A)" — is the most known protocol for agent-to-agent interactions.

>> **Situational instability:** Although the protocol usually is stable, developers still encounter occasional bugs or version mismatches, especially while implementations evolve.

>> **Potential model inflexibility:** MCP's client–server architecture could become limiting as Agentic AI systems evolve toward more decentralized, peer-to-peer, or mesh-like topologies. While MCP standardizes communication between clients and servers effectively today, future multi-agent ecosystems may require more flexible, distributed coordination patterns.

Agent Network Protocol

The *Agent Network Protocol* (ANP) aims to build a decentralized agent collaboration network. Unlike the client–server approach of MCP (see the section on MCP earlier in this chapter), ANP is agent-centric allowing each agent to operate on equal footing. It emphasizes cross-platform interoperability among agents by using decentralized identity (DID) standards and semantic web *linked-data technology* (such as *JSON-LD*). Think of ANP as a kind of internet for AI agents; one that's open, secure, and built to scale massively while these agents become more common in our digital lives.

What sets ANP apart from MCP is its focus on sustained decentralization. This peer-to-peer structure allows agents to establish their own secure identities, negotiate directly with other agents, and form relationships without needing a central authority to mediate: a strong shift toward autonomy, not just for the AI agents themselves, but for the networks that they form.

Technically, this decentralized setup is both ANP's greatest strength and biggest challenge:

>> **It offers resilience and flexibility.** Without a central hub, there's no single point of failure, and the system can grow organically as more agents join.

>> **It's significantly more complex to build and maintain than MCP.** Managing secure identities, routing messages, and discovering other agents in a decentralized system require a level of infrastructure that many organizations can't yet handle.

ANP is an open protocol and still in the early stages of development at the time of writing; it has yet to see widespread enterprise adoption. That's not unusual for a protocol this ambitious. It's laying the groundwork for what could become the foundation of large-scale, autonomous AI networks in the future. But in the meantime, organizations looking to experiment with ANP need to be prepared for a steep technical learning curve and the need to build or integrate new support systems. Table 3-1 gives the ANP pros and cons.

TABLE 3-1   ANP at a Glance: Pros and Cons

Pro	Explanation
Scalability vision	Intended to support networks of billions of AI agents, making it future-ready for the growth of agentic AI.
Decentralized architecture	Reduces reliance on centralized infrastructure and minimizes single points of failure.
Open standards	Community-driven and open source, encouraging transparency, flexibility, and innovation.
Con	**Explanation**
Complex implementation	The decentralized model introduces challenges in identity management, agent discovery, and coordination.
Early stage	ANP is still maturing, with limited real-world deployments and enterprise use cases.
Infrastructure demands	Organizations may need significant technical capabilities to adopt ANP effectively.

Agent2Agent protocol (A2A)

The *Agent-to-Agent protocol*, or A2A, is another open-source protocol developed to meet a very specific need: enabling direct, meaningful communication between autonomous agents which includes discovery of capabilities, task orchestration, state updates, multimodal content exchange, and long-running workflows. As AI systems become more specialized and collaborative, a single agent needs to do more than operate in isolation or simply pull data from a tool. Often, several agents need to work together to negotiate roles, share updates, and coordinate actions. Google developed A2A with this kind of collaboration in mind, and the protocol is now being advanced by a growing open-source community of contributors.

Fostering agent communication

REMEMBER

If MCP is like a phone line between an AI model and a tool, and ANP is a vast decentralized internet for agents, then A2A sits somewhere in between as the Slack channel for chats between agents. (See the sections "Model Context Protocol" and "Agent Network Protocol," earlier in this chapter, for discussion of these protocols.) A2A doesn't aim to reinvent how agents connect across the entire internet, nor does it focus on linking AI tools to models. Instead, it zeroes in on one essential capability: helping autonomous AI agents talk to each other. As Agentic AI grows in complexity and use, I anticipate that protocols such as A2A will become indispensable, even if they remain under the radar for now.A2A focuses tightly on the mechanics of agent-to-agent conversations. The protocol defines clear rules for how agents talk to one another, ensuring their messages follow a structured format, that conversation threads and state context can be tracked across multiple steps, and that agents can effectively share responsibilities within a larger workflow.

This structured approach allows agents that have different roles or specialties to coordinate in complex scenarios. For example, one agent might handle scheduling, another might retrieve data, and a third might verify legal compliance, with each communicating in turn to complete a task together.

Barriers to agent collaboration

Despite the promise of agent communication and collaboration (see the preceding section), A2A hasn't gained the same level of traction in the enterprise world as some other protocols at the time I'm writing this. Here are a couple of A2A's downsides:

>> **Slower industry adoption may limit the size and speed of A2A's developer community.** And a large, responsive community is essential for ironing out bugs and evolving the standards.

>> **A2A makes multi-agent collaboration possible, but it doesn't make it easy.** Managing multiple agents in conversation introduces complexity, especially when agents disagree or when the system needs to maintain state across long interactions. And because agents can have different architectures or models, developers find making sure that they understand one another and respond correctly an ongoing technical challenge.

Table 3-2 gives you an overview of A2A pros and cons.

Agent Communication Protocol

The *Agent Communication Protocol,* or ACP, is an open-standard communication protocol designed to enable AI agents — regardless of framework or deployment environment — to interact directly and collaborate. It is maintained under open governance and underpins the BeeAI ecosystem, allowing agents to discover each other, exchange multimodal messages (text, files, and streams), support long-running workflows, and operate both synchronously and asynchronously.

TABLE 3-2 **A2A at a Glance: Pros and Cons**

Pro	Explanation
Multi-agent focus	Designed specifically for agent-to-agent collaboration, filling a crucial gap other protocols don't directly address.
Workflow support	Enables agents to coordinate on complex, multi-step tasks with specialized roles.
Structured communication	Standardized message formats and interaction patterns potentially make coordination more reliable.

Con	Explanation
Limited adoption	Fewer enterprise implementations mean a smaller ecosystem and slower community growth. But that may change swiftly with the expected uptake in agentic AI adoption over time.
Coordination complexity	Managing conversations, resolving conflicts, and maintaining shared context across agents are technically demanding.
Interoperability challenges	Ensuring different agents can reliably interact across varied models and frameworks remains a significant hurdle.

TECHNICAL STUFF

Although ACP originated in IBM Research's BeeAI project, the ecosystem is growing and invites broad community contribution. The protocol now supports contributions from multiple organizations and open-source developers through GitHub repositories such as i-am-bee/acp at https://github.com/i-am-bee/acp.

Think of ACP like a universal communication system for AI helpers. Just like people speak common languages so that they can understand each other, ACP gives AI agents a standard way to send messages back and forth, no matter who made the agents or what technology they use. As of this writing, Google's Agent-to-Agent (A2A) protocol is better known and more widely discussed, but ACP is an important complementary effort that aims to provide an open, community-governed standard for agent interoperability.

Incorporating internet methods

ACP uses internet-friendly methods (relying on familiar web standards like HTTPS, JSON, and REST APIs) to let AI agents ask questions, share information, and coordinate tasks. It supports multiple conversation styles between agents, including quick back-and-forth exchanges, ongoing discussions where agents can remember what was said before, and streaming continuous data when needed.

ACP is a RESTful application programming interface (API) specification designed to enable seamless, standardized communication between AI agents regardless of their underlying frameworks, platforms, or programming languages. (REST stands for Representational State Transfer, and a RESTful API uses the REST architecture.) As a RESTful API, ACP facilitates a wide range of interaction patterns, including synchronous and asynchronous calls, streaming data exchanges, and stateful or stateless conversations. This extensibility allows users to deploy ACP flexibly across diverse environments, from fully centralized server infrastructures to highly distributed, edge-based systems.

Focusing on precision

ACP offers a highly structured and secure option for organizations that need to manage agent communications with precision. Although ACP may not offer the best fit for open or decentralized networks, it fills a critical niche for enterprises looking to harness AI agents in environments where governance, traceability, and control matter most.

MCP and ACP are complementary standards that help AI systems connect and collaborate. MCP, created by Anthropic, lets AI models and agents access tools, data sources, and external systems. ACP, developed by IBM Research's BeeAI project, builds on similar ideas but focuses on how agents communicate with each other.

ACP isn't a direct extension of MCP, but the two protocols work well together. MCP handles the tool access layer, while ACP manages the conversation layer and together, they make large-scale multi-agent collaboration possible. Specifically, ACP

>> **Helps companies keep close control over how their AI agents communicate,** ensuring that conversations are more reliable and easy to track. Especially important for heavily regulated industries such as healthcare, finance, or government, where privacy and accuracy matter critically.

>> **Facilitates seamless interaction of AI agents created by different developer teams or companies**, even when those agents are built with different tools, programming languages, or deployment environments. In other words, it enables agents developed independently — by different organizations and for different platforms — to communicate and collaborate effectively.

Table 3-3 gives an overview of ACP pros and cons.

TABLE 3-3   **ACP at a Glance: Pros and Cons**

Pro	Explanation
Open and community-driven	Developed as open source by IBM Research's BeeAI project and now community-governed under the Linux Foundation at https://github.com/orgs/i-am-bee/discussions/5.
Flexible deployment	Supports agents in centralized, distributed, and hybrid architectures across cloud, on-prem, and edge environments.
Robust conversation management	Threaded state and structured flows ensure continuity, traceability, and auditability in agent interactions.
Interoperable	Enables communication between heterogeneous AI agents built on different frameworks, platforms, or programming languages.
Con	**Explanation**
Deployment complexity	Scaling multi-agent networks may require careful orchestration and infrastructure design.
Ecosystem maturity	While rapidly evolving, ACP and related tooling are still maturing compared to established communication standards.
Niche focus	Targeted specifically at agent interoperability and AI workflow coordination, not general network messaging.

Building AI Agents

The process of building Agentic AI systems starts with a step that gives the AI a purpose. For example, the system may help manage a busy executive's schedule, run an online storefront, or coordinate a fleet of delivery drones. Whatever the overall mission, the agentic system needs these characteristics:

>> **A method to interpret goals and break them down into smaller, achievable actions:** Unlike simpler forms of AI that rely on direct prompts, Agentic AI reasons through decisions. It weighs options, considers consequences, and chooses what to do next based on logic, past experience, and current conditions. This process of planning and adaptation makes it possible for the AI to function in messy, unpredictable real-world environments, which you can read about in the section "Driving Multi-Agent Coordination and Planning," earlier in this chapter.

>> **A sense of memory that includes an awareness of what has happened before and what matters now:** Agentic systems need both the ability to reason and to act, and that means connecting with tools such as apps, data sources, devices, or even other agents. Developers create these integrations

using software bridges known as application programming interfaces (APIs), or through newer protocols. Once the connections are in place, the AI can send e-mails, update records, make purchases, or carry out any number of other tasks on its own. In the section "Connecting Contextual Awareness and Situational Reasoning," earlier in this chapter, I talk about the core technologies which provide the capabilities of perception, memory, logic, and coordination to create the foundation for collective intelligence in agentic systems.

>> **A capacity to learn from mistakes, recognize patterns, and improve system performance over time:** What makes agentic systems work in practice is a feedback loop. Every action the AI takes produces results, and those results become part of the system's memory. Did the e-mail get a response? Was the hotel reservation confirmed? Did the chosen route avoid traffic? The answers help the agent figure out what works and refine its behavior over time. In more advanced systems, this cycle of action and feedback becomes a form of continuous self-improvement. Flip back to the section "Self-Correcting Continuous Improvement," earlier in the chapter, for an explanation of how Agentic AI systems are designed to learn and adjust their behavior in real-time.

WARNING

Of course, giving an AI that much autonomy comes with serious responsibilities. Developers must build in safeguards to ensure the agent stays within ethical and operational boundaries. That could mean limiting what the AI is allowed to do, requiring human approval for sensitive actions, and/or programming in rules to align with laws and values. Trust and safety become critical when systems are acting independently. See Chapter 10 for a discussion on building agentic systems responsibly.

Choosing technical architecture approaches

Developers can choose from several technical architecture approaches to build Agentic AI systems. For now, consider that the general design and development process follows these steps:

>> **Define the agent's goals clearly** by including specific inputs, desired outputs, constraints, and success metrics.

>> **Analyze current processes** (which you want the agentic system to replace) to understand existing workflows, identify inefficiencies, and determine whether an agentic solution truly fits the need.

>> **Design the target process** by mapping out the desired future workflow, using defined tasks, dependencies, and human intervention points.

Building from scratch

Building an Agentic AI system from scratch provides complete control over its design, data, and operation but demands deep expertise in data management, machine learning, natural language processing, and system integration. This approach involves developing custom algorithms, building and securing data pipelines, implementing memory and reasoning modules, and maintaining continuous optimization processes. It also means handling infrastructure, monitoring, and orchestration tasks yourself. While building Agentic AI from scratch offers maximum flexibility it comes at a steep cost in terms of greater complexity and development time.

REMEMBER

I don't recommend taking the from-scratch approach to agentic AI system development unless you have serious AI developer chops and the time and patience to take on this herculean effort.

Using agentic frameworks

When you're ready to build an agentic AI system, you have two main paths: building from scratch or using an agentic framework. Both can get you to the finish line, but they differ greatly in complexity, speed, and control. Most developers find success in setting up an Agentic AI system by using agentic frameworks. It's widely recommended to take this approach for building intelligent agents because systems constructed this way tend to be more stable, scalable, and easier to maintain than those built entirely from scratch.

These frameworks provide pre-built components and structures that remove much of the underlying complexity in handling things like memory management, task planning, and tool integration. This frees developers to focus on designing the agent's goals and logic instead of reinventing the infrastructure. You can find out more about this framework approach in the section "Framework options," later in this chapter.

Frameworks can offer developers low-level or high-level access to agent development.

>> **Low-level access** gives developers direct access to the underlying mechanisms that control how agents think, decide, and act.

>> **High-level access** in frameworks hides away many of these details — including complex technical details and implementation specifics — behind simplified interfaces and pre-built components.

For developers, this distinction is similar to how a low-level programming language such as C gives you direct memory management control, while a high-level language such as Python handles those details behind the scenes so you can focus on higher-level logic.

Choosing the right framework follows the same kind of design analysis that I describe in the section "Choosing technical architecture approaches," earlier in this chapter. During the agent architecture design phase, developers outline the system in functional layers that include perception, reasoning, memory, and action capabilities, and they design these layers to be modular for easier maintenance and scalability. The final framework choice depends on the specific use case, technical requirements, and desired level of control.

Using AI agent-building platforms

You can use AI agent-building platforms, many of which are no-code or low-code solutions that simplify agent creation through intuitive user interfaces. This development approach is simplified, yes; is it easy? no. You get a better look at this option in the section "Building without coding," later in this chapter.

Agent-building platforms operate quite differently from framework-based approaches I described in the preceding section. Instead of writing code, you use drag-and-drop interfaces, visual workflow builders, and configuration panels. Many of these types of platforms provide ready-made templates for common agent types and tasks. They also integrate with systems such as SharePoint and Salesforce, which helps streamline development and deployment because organizations that already use these systems don't need to transfer or duplicate data. The agents can directly access existing records and processes, speeding up configuration and reducing the risk of integration errors.

Supporting system development, regardless of method

Taking a structured approach — combined with the right technical components — can help you or your company build Agentic AI systems that deliver real business value. (Probably the point of all this, right?)

Consider these supporting phases of system development:

>> **Data gathering and preparation:** Critical phases in which you collect high-quality, relevant data for training and operation. This data must be clean, properly labeled, and well-structured. I know data isn't sexy, but it does act as the seat of power for AI performance. You absolutely can't take any shortcuts on data management when building Agentic AI systems. Check out Chapter 10 for more information about the data that Agentic AI systems need.

>> **Implementation:** Establish the fundamental observe-plan-act-learn cycle that forms the core of agentic behavior. (Some frameworks call this the "reason-and-act" or ReAct loop.) Once this is in place, conduct thorough testing, including:

- Unit testing for individual agents

- Integration testing for multi-agent systems

- Simulating real-world scenarios to evaluate performance and uncover edge cases

- Stress-testing to ensure that the system can handle scalability and resilience requirements under varying load conditions

>> **Deployment:** Start with small, controlled implementations that include real-time monitoring, alerting systems, and fallback mechanisms. This approach reduces risk and helps you detect unexpected agent behavior early.

>> **Continuous iteration and improvement:** Gather user feedback, analyze performance metrics, and refine your agents based on real-world operational results. Agentic systems evolve over time, and continuous improvement is key to keeping them effective and aligned with business goals.

TIP

Treat every agent as a living system. Give it a feedback loop, a monitoring dashboard, and a way to learn from outcomes. That's how you keep it useful and on track.

Building without coding

In a nutshell, building without coding means building Agentic AI agents by using low-code or no-code AI agent-building platforms. These platforms aim to take a democratized approach by removing traditional technical barriers through the use of visual interfaces and pre-built components. The platforms are designed to assist non-technical individuals and businesses to create sophisticated AI-driven assistants without extensive programming knowledge. In theory, at least, they make AI accessible to a much broader audience.

Recognizing both benefits and drawbacks

The advantages of AI agent-building platforms often go beyond ease of use. They can dramatically reduce development time and typically provide built-in integrations with popular business systems that can handle much of the complex back-end infrastructure automatically. Most platforms include analytics and monitoring tools that can track agent performance and user interactions without the need to build custom dashboards.

However, these platforms also have limitations. You may have fewer customization options compared to building from scratch, and if you have highly specific requirements, you might find yourself overly constrained by platform capabilities. You can also face concerns about vendor lock-in and ongoing subscription costs, especially while usage scales.

Looking over types of platforms

The choice between AI agent-building platforms typically depends on existing technology infrastructure, specific use-case requirements, budget considerations, and the level of customization you need.

TIP

If you want to go the building-platform route for agentic system development, check to see whether the AI agent-building platforms that you're considering will give you a free trial period to help you determine which platform serves your needs best.

Here are a few platform examples:

>> **Medium advanced platforms such as MindStudio** offer powerful visual builders that make building complex AI workflows using combinations of language, image, and voice models less complicated. An extra plus is that you can tend the agentic system that you build with custom code when you need more advanced control.

>> **Google's Dialogflow platform,** on the other hand, focuses heavily on natural language understanding and conversation design. Dialogflow provides integration with Google Assistant, websites, Slack, Facebook Messenger, Teams, X (formerly Twitter), and many other platforms. Companies that are already invested in Google's ecosystem often find Dialogflow a natural fit.

And Microsoft offers a range of platforms that have varying capabilities:

>> **Microsoft Bot Framework:** Takes a more developer-friendly approach while still offering low-code capabilities. Companies that use Microsoft Office 365 or Azure services often find Bot Framework or Copilot Studio (see the following bullet) a more natural and familiar fit for development.

>> **Copilot Studio:** A no-code AI agent builder designed for creating AI-driven conversational or task-driven agents. It seamlessly integrates with other Microsoft services as part of the Power Platform, making it particularly attractive for organizations that already use Microsoft's tools.

>> **IBM watsonx Assistant:** Has feature-rich capabilities and strong support. Companies that require extensive AI capabilities that come with enterprise-grade support might prefer IBM's watsonx Assistant, despite its potentially steeper learning curve.

Building with coding

You can build Agentic AI agents and systems from scratch (by coding) if you have the skills to do so. Honestly, very few people do. But if you're feeling froggy and want to jump right in anyway, Python is currently the most common programming language used in building agentic AI systems and agents.

From-scratch options (or not)

Python's popularity as the programming language for AI stems from its simplicity, readability, and a rich ecosystem of libraries, such as TensorFlow, PyTorch, spaCy, LangChain, and scikit-learn. These libraries make building, deploying, and scaling complex AI systems — including agentic architectures and frameworks such as LangChain and AutoGen. . . — both accessible and efficient. Further, Python's extensive community and ongoing support make it highly adaptable for the rapidly evolving requirements of agent-oriented AI applications. However, agentic developers also use the programming languages C++ and Java, while Julia, and R are less common.

REMEMBER

As I mention in the section "Using agentic frameworks," earlier in this chapter, the broader AI developer community, including major research and industry players — including Microsoft, OpenAI, Anthropic, and the open-source LangChain/AutoGen communities — all recommend that you use agentic frameworks to build agents (see the following section), rather than the quite nightmarish DIY-from-scratch approach.

Framework options

Agentic system-building frameworks provide pre-built components and structures to make the process easier (though still not easy). In this context, "frameworks" can range from developer toolkits (LangChain, Semantic Kernel) to runtime orchestration systems (AutoGen, CrewAI). And you find that most are specialized frameworks. For example, CrewAI specializes in multi-agent systems in which different agents collaborate on tasks, while BabyAGI offers a simpler yet powerful framework for autonomous task execution. Table 3-4 gives you a look at a few framework choices.

TABLE 3-4   **Agentic System-Building Frameworks**

Framework	Description	Features
LangChain	A Python-based framework popular among developers new to Agentic AI.	Provides granular control for building LLM-powered agents Has extensive documentation and strong community support Connects LLMs with APIs and data sources
LangGraph	An open-source Python framework for advanced AI agent applications and complex workflows.	Provides reliability and controllability Offers straightforward, modular architecture Supports conversational workflows and complex state management
Microsoft Semantic Kernel	A lightweight, open-source software development kit (available in C# and Python) built by Microsoft for integrating AI models into enterprise systems.	Strong enterprise integration capabilities Modular and extensible architecture Works well with Azure OpenAI, OpenAI, and Hugging Face models
LlamaIndex	A framework optimized for knowledge-heavy applications that use context-augmented agents and LLMs.	Offers a flexible framework for building knowledge assistants Connects specialized agents to enterprise or document data Includes strong support for retrieval-augmented generation (RAG)
AutoGen	A multi-agent orchestration framework developed by Microsoft that provides runtime capabilities for creating enterprise-ready AI solutions.	Aligns with Microsoft Semantic Kernel Enables multiple AI agents to work together conversationally Supports human-in-the-loop and autonomous modes
CrewAI	A collaborative multi-agent framework that makes it easier to build and manage teams of specialized AI agents working toward shared goals.	Focuses on coordination among multiple agents Includes built-in task assignment and orchestration Supports natural language collaboration between agents and humans

Chapter **4**

Interacting with Agentic AI

Y ou are here — in the early stages of a new paradigm where you need to direct autonomous agents capable of independent reasoning, planning, and execution. It's a managerial position for which no one trained and, frankly, few applied. Nonetheless, Agentic artificial intelligence (AI) is rapidly emerging, just like Generative AI (GenAI) did a few years ago. Don't worry; you don't have to waste any newly developed prompting skills because GenAI is here to stay. Much of what you may already know about crafting great prompts can apply to directing Agentic AI, too.

The AI world is changing, and new AI agents and agentic systems can create real problems when they're deployed without strong oversight, clear boundaries, and well-trained humans in the loop. As AI agents become more capable of handling complex, multi-step tasks with minimal oversight, both developers and users must develop new skills in delegation, boundary-setting, and collaborative problem-solving.

Effective human–agent interaction requires not just technical understanding, but also a nuanced appreciation for when to intervene, when to step back, and how to maintain meaningful control over a growing number of increasingly autonomous systems.

Meanwhile, take a deep breath and brace yourself for the frontier that's coming your way. This chapter helps you see how human-AI interaction patterns emerge across different contexts — from individual productivity to enterprise workflows — and what the patterns reveal about the future of human-AI collaboration.

Mistaking AI as a Colleague Creates Errors

Even though I'm an AI enthusiast and a cheerleader for progress (and change, in general), I cringe every time I hear someone say that they "collaborate" or "partner" with AI. I cringe even harder when I find those words in an AI's output. It feels *wrong* (by which I mean incorrect, not morally suspect) and ridiculous — like outrageous marketing speak.

So why do I cringe? Because I see using the terms *collaborate* and *partner* as willful misdirection and inaccurate descriptions of what actually happens. AI is a tool. People don't collaborate with tools. They don't partner with tools. They make tools. They use tools. They command tools. That's it. And that's the whole relationship humans have with AI.

In my opinion, equating a tool — no matter how smart it appears — with the intelligence, attributes, or worth of a human is a huge mistake. And that mistake may make you much more likely to believe and trust AI because you forget that it's a tool.

TIP

Recognizing AI as a tool reinforces a mindset that compels you to figure out how to control its activity and use it responsibly.

Establishing the AI-as-tool mindset

If you remember that AI is a tool, you routinely check the AI's output as a matter of quality control and not as a basis for negotiation or debate with an equal. You need to have and act on this AI-as-tool mindset because (as my dearly departed dad would put it) AI "can mess up a can of red worms." This colorful phrase simply means that AI can overly complicate and mess up almost anything, even something rudimentary and mundane. For example, when it decides to summarize a three-sentence e-mail, insists that Tuesday comes after Friday, or recommends that your cat needs a project plan.

Here are other reasons to push back on the idea that AI is equal to (or better than) human intelligence and talent:

>> **If you believe AI is as smart as (or smarter than) you, you can become dependent on AI and lose your own abilities.** AI does have direct access to a lot of data and can come up with information quickly. But any dependence humans have on AI's information doesn't happen because AI is smarter, but because people outsource their thinking and remembering to it.

If you don't believe me, tell me this: What phone number (other than your own) do you remember? Most people have outsourced memorizing phone numbers to smartphones. How much more do you think you might outsource to AI if you aren't careful?

>> **It gives employers false confidence in the wholesale replacement of people with AI.** When business leaders start seeing AI as a cheaper equivalent to skilled workers, they often overcut their workforce only to discover later that AI can't truly replace the judgment, creativity, or adaptability of a human.

Jobs inevitably evolve, as they always do with new technologies, but people aren't interchangeable with machines. Many human roles will endure — transformed by AI, rather than erased — and entirely new roles for humans will emerge alongside them. Meanwhile, employers lay off huge swaths of people only to recall many of them a short time later.

REMEMBER

I make these points to explain why I encourage you to think of this transition from interacting with GenAI assistants to working with Agentic AI systems as a progression from prompting to *directing*, a far more accurate description of interacting with Agentic AI. You direct, instruct, and manage agents. If you do any collaborating, you do it with other humans.

Discovering how to direct Agentic AI

With the correct mindset in place (see the preceding section), consider how you can go about directing and instructing your AI tools to best effect.

You most often encounter Agentic AI through natural language interactions, either in text or by voice. You use an interface to delegate complex, multi-step tasks, rather than just ask simple questions. Unlike traditional GenAI interactions that require precise prompts or keyword use, agentic systems can understand high-level goals such as *help me plan a marketing campaign.* The system then breaks down that goal into constituent tasks on its own.

REMEMBER

Essentially, your interaction with AI systems shifts from micromanaging each step to defining outcomes and letting the AI determine the path. The process, tools, and data involved have these characteristics:

>> **Integrated tool ecosystems:** AI agents connect to existing software, websites, databases, and digital tools via APIs or protocols like MCP (as I explained in Chapter 3). These ecosystems may allow AI agents access to calendars, e-mail systems, databases, or specialized applications, but always under user-defined permissions, constraints, and oversight. Agents do not have unrestricted or autonomous access; their actions are mediated through the integrations you approve.

>> **Access through multiple inputs:** You can also interact with Agentic AI through multiple input methods simultaneously — such as text, voice, images, or structured data. The agent or system maintains context across these different *modalities* (input forms) so they can synthesize the combined information and choose the appropriate next action.

>> **Ongoing dialogues with the ecosystem:** Your work with the agentic system isn't a one-shot interaction. You engage in a continuing dialogue to refine objectives, give or withhold permissions for agent's proposed actions, and provide feedback to guide the agent's next steps. This iterative loop can feel conversational, but the interaction remains fundamentally between a human and a tool.

>> **Changes in the level of interaction:** While AI agents become more autonomous, your interactions with them increasingly involve setting boundaries, monitoring progress, and intervening when necessary. Over time, you can figure out how to provide clearer direction and constraints up front and develop more advanced skills in spotting when an agent might be going off-track.

Comparing Context Engineering to Prompt Engineering

While AI becomes increasingly sophisticated, the ways that you interact with and guide these systems also evolve. Two key practices in this domain are *prompt engineering* and *context engineering.* Prompt engineering focuses on designing clear, effective instructions that steer a model's immediate output. Context engineering, by contrast, shapes the broader environment the model operates within, including its goals, constraints, tools, data sources, and memory.

This environment is essential for Agentic AI systems to function. Although these practices may seem similar, they operate at different levels of complexity and aim to achieve distinct outcomes. Prompt engineering influences what the AI does in a single step, while context engineering determines how agents behave across many steps in a larger workflow.

In essence, prompt engineering involves getting the immediate job done with a GenAI model, while context engineering involves setting up an Agentic AI agent or system to effectively and consistently pursue more complex, ongoing objectives in a dynamic environment. Prompt engineering may be a component within a larger context-engineered system to provide the agent with prompts or instructions within its rich context, but context engineering encompasses a broader and deeper approach to AI system design and operation.

Seeing why you need both practices

Contrary to popular belief, context engineering won't replace prompt engineering — at least, not entirely. Consider these factors:

>> GenAI isn't going away, and those models and applications continue to need prompting. GenAI can even serve as the interface for an Agentic AI system or as the coordinating agent in a multi-agent workflow. Although the GenAI can handle these coordinating and interface tasks quite easily, it can do so only after you prompt it.

>> Creative souls such as musicians, artists, writers, problem-solvers, visionaries, and innovators want more granular control over how an AI tool manifests their ideas or reflects their thinking. Agentic AI probably can't satisfy this need for them.

Autonomy and orchestration are Agentic AI's strengths; precision control is not. Granular creative control requires more prompting, not less. If you already read Chapter 3, then after reading this section, you probably immediately thought of the difference between high-level and low-level access in agentic frameworks. If so, then you're already making the connection: the more control you want, the lower level and more hands-on your instructions must be.

This continuing need for GenAI means that your efforts to figure out how to prompt successfully are not wasted. Keep developing that skill because you will improve how you express goals, constraints, and instructions with clarity and structure over time. That mindset shift helps you articulate tasks in a way that aligns with how agentic systems interpret information. In turn, your commands and stated goals become more precise and effective.

TIP

You also need to get a solid grasp of context engineering because Agentic AI requires more than a few words of input. It also needs a well-structured environment, clear goals, defined constraints, and enough contextual information to manage the many steps involved in delivering the outcome you want. Even the most demanding creative minds (perhaps that's you) can likely find Agentic AI helpful in running the practical side of their businesses or jobs. For example, Agentic AI systems can automate and manage invoices, schedule exhibitions, interact with buyers or supervisors, and handle other necessary but less creative tasks.

Understanding the fundamental differences in engineering methods

The key distinction between prompt and context engineering practices lies in scope and duration. Prompt engineering is like having a really good conversation with an expert consultant. You need to ask clear and specific questions to get valuable answers, but each conversation is essentially self-contained. Context engineering is more like hiring and onboarding a new employee. You're defining goals, tools, rules, and workflows thereby creating an environment that guides the system's behavior over many steps and over time.

REMEMBER

Both disciplines require deep understanding of AI capabilities and limitations, but they operate at different scales and timeframes. A well-engineered prompt might generate a perfect e-mail or analysis in seconds, while a well-engineered context might enable an AI agent to successfully manage complex, ongoing tasks over weeks or months.

This table shows the knowledge of AI systems needed for each practice.

Prompt Engineering Requires	Context Engineering Requires
Understanding how language models process and respond to text	Understanding how autonomous systems make decisions
Figuring out what kinds of instructions work most effectively	Designing robust workflows
Developing skills in clear communication	Achieving balance between giving an agent enough freedom to be useful and maintaining appropriate constraints and oversight to prevent problems

While AI technology continues to evolve, AI users need to have both context engineering and prompt engineering skills to use for different applications. Organizations that implement GenAI for content creation, analysis, and creative tasks need to rely heavily on prompt engineering expertise. Meanwhile, businesses that deploy autonomous AI agents for operations, customer service, and process automation need strong context engineering capabilities. Figure 4-1 compares both methods so that you can easily see their differences and individual strengths.

Prompt Engineering	🤖 **Context Engineering**
For Generative AI Systems	*For Agentic AI Systems*
→ Interaction Model Single-turn conversations where each prompt is a complete, self-contained request	**→ Interaction Model** Ongoing autonomous operation with persistent memory and continuous decision-making
→ Time Horizon Immediate responses focused on one-time outputs like text images, or analysis	**→ Time Horizon** Long-term operation spanning hours, days, or weeks with evolving objectives
→ Primary Skills Clear communication, instruction crafting, understanding model behaviors and limitations	**→ Primary Skills** Systems thinking, workflow design, constraint definition, tool integration
→ Key Techniques Few-shot learning, chain-of-thought prompting, role specification, output format control	**→ Key Techniques** Environment setup, tool access configuration, goal hierarchy design, safety constraints
→ Example Use Cases Content generation, language translation, code writing, creative writing, data analysis	**→ Example Use Cases** Customer service automation, research monitoring, task scheduling, process optimization
Best For: Creative tasks, content creation, one-off analysis, and situations requiring human-like text generation	**Best For:** Repetitive tasks, complex workflows, autonomous decision-making, and systems requiring ongoing operation

FIGURE 4-1: A side-by-side comparison of prompt engineering and context engineering.

Generated with AI using Claude AI

Examining the basics of context engineering

Computer scientists and AI practitioners increasingly use the term context engineering to describe the discipline of designing, structuring, and managing the contextual information that an AI system relies on. But I like to put the description in more practical terms: Context engineering is the process of building and managing the comprehensive understanding or operational environment for an AI agent. It's the work that humans do to provide depth to AI's understanding of its directions. Context engineers provide agentic systems with a rich, dynamic, and persistent grouping of data, tools, and rules that helps the system grasp its purpose, track its history, and interact effectively with the external world.

Think of context engineering like setting up a highly capable human assistant with the resources and tools that they need for a long-term project. You don't just tell them, "Write a report," and then shoo them out the door. Instead, you brief them on the project's objectives, provide access to relevant documents and data, introduce them to key team members, explain the assistant's specific role, and clarify the tools that the assistant can use. This entire body of information, which continuously evolves while the project progresses, provides an example of the context that an AI agent needs.

Providing comprehensive awareness

AI agents need comprehensive contextual awareness to function effectively, including not just the immediate query, but also relevant historical information, domain knowledge, available tools, and operational constraints. Context engineering is the method (and discipline) that ensures agents have access to all necessary information in a structured, accessible format.

Context engineering involves a process of careful design and continuous refinement, rather than a single setup. You typically need to create layered information architectures If you're a context engineer, you follow this set of general steps:

1. **Establish the agent's core knowledge.**

    Engineers typically use Retrieval-Augmented Generation (RAG) or other systems that allow the agent to access relevant documents, databases, or knowledge bases.

2. **Implement memory systems.**

    The memory layers track conversation history, user preferences, decisions, and previous outcomes, allowing the system to maintain continuity across tasks.

3. **Add a tool orchestration layer.**

   This layer defines the agent's capabilities, available tools, and decision-making policies that determine when and how those tools can be used.

4. **Implement dynamic context-selection algorithms.**

   These algorithms determine which information is most relevant for each specific interaction. This layer is aimed at ensuring the agent isn't overwhelmed by unnecessary data.

During development, the engineer embeds guardrails and constraints to keep the agent operating within safe and well-defined boundaries. They also design the system to scale so that it can accommodate context that can grow quite large over extended interactions. Do keep in mind that Agentic AI systems are new and still evolving so these methods will change and improve over time.

Interconnecting system components

Take a look at the following explanations of some of the techniques and components that context engineers must interconnect so that they can work together to create a comprehensive information environment for AI agents. Following along with the steps for designing the AI system's architecture (see the preceding section), context engineers can incorporate these components and techniques:

» **Retrieval-Augmented Generation (RAG):** Frequently used as a foundational component. This base layer or core building block allows agents to access current, domain-specific information that isn't stored in their training data.

» **Memory systems:** Enable agents to maintain context across multiple interactions, including both short-term memory for ongoing conversations and long-term memory for persistent user preferences, past decisions, and learned patterns. Memory systems help agents build continuity and avoid repeating mistakes or asking for previously provided information.

» **Tool integration:** Provides agents with extra abilities. Context engineering involves specifying the available tools, the rules governing their use, and the procedures for interpreting tool outputs. Tools might include databases, application programming interfaces (APIs), calculators, web search capabilities, or domain-specific software functions.

» **Dynamic context assembly:** Perhaps the most sophisticated aspect of Agentic AI system architecture. The system must intelligently determine which contextual elements are relevant for each specific task or query.

» **Error learning and adaptation:** Typically through machine learning. This advanced technique helps agents learn from mistakes and improve performance over time.

Context engineering is vital for Agentic AI, but carrying it out effectively involves several challenges, including managing dynamic information and sometimes inconsistent information, as well as the current limitations of AI models in reasoning, memory, and long-horizon tasks. Addressing these issues requires advancements in large language model (LLM) architectures, more sophisticated context management methods, stronger filtering and relevance-selection systems, and careful ethical and safety design. Table 4-1 offers a quick look at the challenges involved in context engineering for Agentic AI systems.

TABLE 4-1 **Challenges in Context Engineering for Agentic AI**

Context Element	Challenge	Impact
Context window limitations	LLMs have finite context capacity, often exceeded by Agentic AI's need for long-term memory and complex goal tracking.	Information loss, poor recall, decreased performance.
Maintaining contextual coherence	Context can become fragmented or contradictory over time, making it hard to maintain consistent understanding.	AI forgets instructions, misinterprets situations, pursues irrelevant goals.
Handling ambiguity and nuance	AI agents can find extracting true intent from ambiguous human language and real-world situations difficult.	Misunderstandings, incorrect actions, lack of robustness.
Dynamic context adaptation	Context constantly changes; AI needs to rapidly adapt without losing core objectives.	Slow reactions, inability to pivot, reliance on outdated info.
Information overload and irrelevance	AI agents may find filtering vast amounts of data to find pertinent context challenging.	High computational cost, *hallucinations* (which present fabricated or unsupported information as factual), diluted signals.
Bias and misinformation propagation	Context can contain biases or misinformation, which the AI might amplify.	Unfair outcomes, false information, eroded trust.
Cost and latency of context processing	Processing large contexts is computationally expensive and time-consuming.	Slow responsiveness, increased costs, limited task complexity.
Ethical considerations	Deciding what context to include or prioritize has ethical implications (privacy, fairness).	Privacy breaches, biased decisions, difficulty in accountability.

Augmenting context engineering with prompt engineering

As I introduced in the section "Comparing Context Engineering to Prompt Engineering," earlier in the chapter, prompt engineering is simply the skill of shaping your instructions so that a GenAI model gives you better results. This practice ranges from writing clear questions to using more advanced techniques, such as asking the model to show its reasoning (chain-of-thought), giving it a few examples to follow (few-shot prompting), or assigning it a role to guide its tone and expertise.

REMEMBER

The relationship between prompt engineering and context engineering is deeply symbiotic. Context engineering defines the information architecture, tools, constraints, and environment an agent works within, while prompt engineering provides clear, structured instructions that help the language model interpret and act within that environment.

If you want to find out more about prompting methods and techniques, read my books *Generative AI for Dummies* and *ChatGPT for Dummies* (Wiley). Also, check out my LinkedIn Learning course titled "Become a Generative AI Power Prompter and Content Creator," at `https://www.linkedin.com/learning/become-a-generative-ai-power-prompter-and-content-designer/become-a-genai-power-user` which offers a lot of hands-on exercises to help you develop your prompting skills.

Incorporating prompt engineering in Agentic AI

Rather than getting replaced by context engineering, prompt engineering remains a crucial component within the larger context-engineering framework. In Agentic AI systems, prompts serve as the interface between the engineered context and the LLM's processing capabilities. Because of this inherent relationship, prompt engineering acts as a subset or a component within broader context-engineering systems.

Prompt engineering provides the basic instruction templates that define how AI agents and agentic systems should interpret and use their contextual information. Here are a couple of specific functions of prompts in an Agentic AI system:

>> **Putting the focus on the right information:** When building Retrieval-Augmented Generation (RAG) systems, for example, well-engineered prompts determine how retrieved information gets integrated with user queries. A poorly constructed prompt might lead an agent to ignore relevant retrieved documents or misinterpret their significance, while a well-engineered prompt ensures the agent weighs and synthesizes information appropriately.

>> **Establishing the agent's foundational instruction set:** The *system prompt* is the foundational instruction set that defines an agent's role and behavior and is the most critical application of prompt engineering within context engineering. The system prompt must clearly establish the agent's purpose, communication style, decision-making framework, and operational boundaries. It serves as the agent's interpretive lens through which all contextual information gets processed.

The challenge for any agentic system doesn't involve just gathering relevant context, but also structuring that context so the model can use it effectively. That structuring is complex: Context-engineering systems pull information from many sources — such as memory modules, tool outputs, retrieved documents, and real-time data streams — and then filter and prioritize that information to create a coherent, usable context for the model.

For instance, when an agent accesses both historical conversation data and current database information, prompt engineering principles guide how this information gets organized within the input. Should historical context come first or last? How should conflicting information be flagged? How can the prompt structure help the model prioritize different types of information? These decisions on how to design the prompt directly impact agent performance.

Prompting and tool integration

In agentic systems, prompt engineering becomes essential for tool integration, as well as for gathering information. When an agent needs to decide whether to search a database, call an application programming interface (API), or perform a calculation, the prompts must clearly establish decision criteria and usage patterns. The instructions for how to interpret tool outputs and integrate them into ongoing processing also rely heavily on prompt engineering techniques.

Here are a couple of examples:

>> **A customer service agent that has access to order databases, inventory systems, and policy documents:** Prompt engineering determines how the agent evaluates which tools to use for different query types, how to combine information from multiple tools, and how to present findings to customers in helpful ways.

>> **Implementation of error handling techniques used by Agentic AI systems:** When agents encounter failures or produce incorrect outputs, carefully designed prompts can guide them to evaluate the issue and adjust their approach in the moment and within the task. These prompts may encourage the agent to reflect on unsuccessful actions, reconsider earlier steps, or follow instructions for incorporating error feedback into its next decision.

Maintaining agent behavior

While context engineering systems operate over extended periods, maintaining consistent agent behavior becomes more challenging. Prompt engineering provides the structural framework for this consistency. By establishing clear behavioral guidelines, communication patterns, and reasoning approaches in the foundational prompts, context engineers ensure agents maintain coherent personalities and approaches, even while they access new information and encounter novel situations.

Without skilled prompt engineering, even the most sophisticated context engineering systems may fail to produce reliable results because the LLM can't effectively interpret or use the provided context. Conversely, prompt engineering alone, meaning without the broader information management and tool integration that context engineering provides, limits AI systems to relatively simple, isolated interactions.

Evolving Voice, Intent, and Semantic Interface Design

Agentic AI interface design is rapidly evolving across multiple modalities (voice, text, and intent-driven interfaces), each bringing unique advantages and challenges for human-AI interaction.

For example, voice processing, no longer just a convenience, is fast becoming a foundational interface for Agentic AI, especially in mobile, embedded, and hands-free environments. Although early voice assistants such as Apple's Siri and Amazon's Alexa relied on fixed commands and linear conversations, Agentic AI demands functionality that's far more adaptive.

In practice, these interface types don't operate in isolation. A fully Agentic AI system often blends all three modalities of voice, intent, and text. For example, it may use voice to receive casual instructions, intent inference to translate that into action plans, and text understanding to navigate complexity and ambiguity.

Voicing your intent with AI

Modern voice interfaces for agentic systems are evolving to be increasingly context-aware, memory-enabled, and multi-turn interactions capable. These improvements mean that instead of rephrasing or repeating commands when a system fails to understand your intent, you can speak naturally and even reference prior parts of a conversation. For example, you may say, "Can you e-mail that report we talked about yesterday?" Agentic AI interprets this by using the conversational context and stored information from prior interactions, not just the literal words spoken in the moment.

In agentic environments, voice becomes more than a control surface; it is evolving to more closely mimic human-to-human conversations. Specifically

» The agent may proactively speak to alert you of an anomaly or request clarification when it doesn't understand your intent. This bidirectional use of voice starts to resemble co-working, rather than tool use. (But it's still you using a very good tool in a smart and natural way; see the section "Mistaking AI as a Colleague Creates Errors" earlier in the chapter.)

» Intent-based interfaces go beyond understanding the what of your request to interpreting the why of it. This interaction feels like a conversation with a living entity or co-worker. Instead of parsing commands literally, these systems infer your goal and determine the best course of action to fulfill it, even if you give incomplete or ambiguous instructions.

REMEMBER

In Agentic AI, the capability to infer and interpret is crucial. An agent needs to make autonomous decisions on your behalf, which means it must reliably recognize your intent, even when you don't explicitly state it. For example, you might say, "I need to reschedule tomorrow's calls," and the agent analyzes your calendar, alerts the attendees, finds alternative time slots, and updates everyone accordingly without your having to dictate each of those steps.

These voice interfaces are becoming increasingly adaptive and context-aware, rather than strictly rule-based. They can pick and choose from your preferences, your past behavior, contextual signals such as location and time of day to clarify your intent and choose the necessary actions.

Interpreting meaning with semantic interfaces

Semantic (text) interfaces enable Agentic AI to understand and process the meaning behind a variety of inputs. These agents look beyond syntax and keywords to grasp relationships, metaphors, ambiguity, and domain-specific language.

Semantic parsing (which converts natural language into a structured representation of its meaning) plays an especially important role in multimodal agentic environments in which instructions may come from voice, visuals, gestures, or combinations thereof. For example, you might say, "Move this there" while pointing to objects on a screen. The AI must resolve what *this* and *there* mean based on context and shared knowledge. As of this writing, agents' ability to resolve such issues is at a very early stage and brittle.

Semantic interfaces also enable agents to work across domains and systems. For example, the interface can connect meaning from a project management app, an e-mail thread, a voice conversation, and a whiteboard sketch into a unified internal model.

Rising Hyper-Real AI Avatars

The evolution in interface design that uses AI avatars is about both technology and relationship design: building interfaces that allow AI agents not only to serve you meaningfully, but also to appear more human. They can be designed to anticipate your needs. And all the while, the avatars adapt to your changing goals, lasting habits, and evolving environments.

And that brings us to 2D and 3D digital human avatars as one of the newest options in Agentic AI interfaces. This isn't a future prediction; several brands, such as Salesforce, already use these interfaces. Examples of vendors that offer avatar interfaces include Soul Machines, Synthesia, Nvidia, Hour One, Replika, and Ravatar. While these avatars exist, widespread enterprise adoption is still emerging. Many implementations are customer service chatbots or marketing demos rather than full Agentic AI systems that deeply anticipate needs and adapt over time.

REMEMBER

A digital avatar isn't just a visual layer. It's the front-end for a complex multimodal interface. When paired with an Agentic AI that understands voice tone, semantic intent, and contextual meaning, the avatar becomes an expressive, adaptive, interactive, and situationally aware entity.

Yet these early avatars aren't very lifelike. Their movements look stiff, their eyes don't always focus on you, and there's no real sense that anything meaningful is behind the face. That makes it tough for users to feel a genuine connection with the AI that the avatar represents.

WARNING

Even so, people have to watch out for *anthropomorphism* (attributing human traits to nonhuman entities) and misplaced trust with these digital humans. Lifelike avatars may increase user engagement, but they also risk over-humanizing the AI. When a digital face nods or smiles, people may ascribe understanding or empathy to the agent that doesn't actually exist. Designers must balance emotional resonance with transparency. Users need to be fully informed when they're speaking with a machine, even if that machine looks and sounds convincingly human.

Personalizing Workflows

Agentic AI systems are redefining how people personalize their workflows, not by tweaking templates or static rules, but by actively shaping experiences around the individual. As a result, workflows no longer follow one-size-fits-all logic. They can shift in real time to meet you where you are, anticipate what you need, and adjust course as circumstances evolve.

At the heart of workflow personalization is memory. Not just short-term recall, but persistent recall of evolving user behavior, preferences, and intent. For example, a content marketer working in a campaign management platform might find that, over time, the agent begins suggesting subject lines that reflect that marketer's preferred tone, often-used layout structures, and even creative formats that are tailored to their audience engagement history. The agent wasn't explicitly programmed with these preferences; it adapted to them by observing choices, measuring outcomes, and adjusting its behavior accordingly.

Altogether, Agentic AI transforms workflows from static pipelines into living systems — contextual, dynamic, and deeply personal. By understanding who users are, how they work, and what outcomes they seek, these agents can do more than automate. They adapt and evolve to create more intuitive, comfortable, and highly effective workflows.

Taking informed actions

Agent personalization isn't limited to your tastes or preferences, and it far exceeds GenAI capabilities. For example, he difference between GenAI integrated with help desk software and Agentic AI systems lies in the extension of decision-making to situational, autonomous action.

Consider a customer support environment in which help desk tickets flood in unpredictably. A conventional GenAI assistant might help draft replies or suggest routing rules, but it still relies on predefined workflows. An Agentic AI system, by contrast, continuously evaluates each ticket in context, such as by customer importance, issue urgency, available agent skills, prior interactions, and even the emotional tone of the message. It then dynamically routes the ticket accordingly. This autonomy allows the system to adapt moment by moment, rather than simply following static rules.

REMEMBER

The workflow for Agentic AI in customer service isn't a set process; it shifts second by second to optimize outcomes, reduce *churn* (the loss of customers), and preserve human bandwidth for the most sensitive or complex interactions.

Adapting to (and for) user's roles

With Agentic AI, workflows are adapted by role. Take a hospital environment, for example:

>> **A nurse accessing a patient dashboard** might see a prioritized view of medication schedules, vital readings, and care coordination notes.

>> **A physician accessing the same patient record** sees the interface designed for physicians where the agent might show suggested next steps based on test results interpretation, prognosis tools, and prescription authority.

In other words, the same underlying data is repurposed for different cognitive workflows, and each representation is personalized to the professional's scope of responsibility and authority. This is similar to role-based dashboards offered in other types of software but far more dynamic.

Personalization by user's role also extends to knowledge work. Imagine an enterprise planning scenario where an executive needs to draft a strategic roadmap. Instead of gathering documents and leaving it at that, an Agentic AI could assemble prior plans authored by the same executive, blend in the latest performance metrics, and incorporate board feedback from past quarters. The Agentic AI produces a highly relevant, contextualized draft of a plan tailored not just to the task, but to the decision-maker behind it.

Multimodal interaction adds yet another layer of customization to interactions with Agentic AI. While still in their infancy, the goal is that when agents appear in 2D or 3D digital human form, they can tune speech pace, body language, and interaction style to better engage with specific users. For instance, a digital avatar guiding new hires through onboarding might present instructions in simplified language for early-career staff, while offering more technical details to experienced professionals. This level of adaptation fosters engagement, trust, and accessibility in ways that static interfaces may struggle to deliver.

Tailoring output based on user feedback or co-agent needs

Unlike traditional automation, Agentic AI is designed to be inherently self-improving. It seeks feedback from you and other users, monitors outcomes, and iteratively refines its workflows. In a financial setting, an Agentic AI might initially offer budget recommendations that feel misaligned with your comfort zone. But with feedback, either explicit (rejections) or implicit (ignored suggestions), the AI adapts and recalibrates. Over time, it produces financial guidance that aligns more closely with personal goals, values, and habits, whether that's aggressive saving, ethical investing, or optimizing for changing family needs.

Crucially, these agents don't operate in silos. In multi-agent systems, personalization can occur across the network. One agent finalizing a supplier contract might tailor its outputs for another agent managing logistics, ensuring that the multi-agent system also optimizes downstream actions for efficiency and compliance. In other words, customization doesn't stop at the individual level. It can extend across workflows, systems, and even organizational layers.

In creative domains, this adaptability manifests in more collaborative, co-pilot-style behavior. A software engineer who uses an AI coding digital assistant may one day find that the agent not only predicts what code to write next, but also remembers project-specific conventions, flags inconsistencies, or initiates test runs based on prior patterns. Rather than providing just code generation in the way that GenAI tools might, newly evolving Agentic AI designs offer personalized problem-solving embedded in the development lifecycle.

Shifting from Apps to Agents

The emergence of Agentic AI marks the beginning of a massive transformation in how people will interact with software and the web. Unlike traditional applications that passively wait for user input to initiate an action, Agentic AI systems automate processes to actively (and sometimes proactively) deliver an outcome to the user. Think of a smart dog that waits for commands from its owner before acting, versus a highly trained military dog that instantly adapts to the current scenario and proactively acts accordingly. The first dog responds; the second dog anticipates, plans, and executes.

This paradigm shift from apps to agents is in the very early stages at the time of writing, but it's more than theoretical. The abilities of Agentic AI already reshape how people think about software interfaces, productivity, e-commerce, and even the architecture of the app itself. Software such as Adobe Acrobat or Microsoft Office make a point of offering AI help whenever you interact with a document or begin a task in that program.

The difference between traditional applications and Agentic AI systems lies in agency. Traditional applications have none, and the agentic systems have some. In this instance, *agency* means the capability of the system to act on its own based on goals rather than commands. Agency — or lack thereof — manifests like this:

>> **Traditional apps essentially offer very specific functions, and only those functions.** For example, a calendar app holds events, a CRM app stores contacts and interactions, and a travel app lets you view and book available flights. You provide all the intent, context, and decision-making logic behind the actions that you command an app to take.

>> **Agentic AI systems are goal oriented.** In other words, an Agentic AI system is on a mission. For example, you can establish the mission by giving it a high-level task in a typed or spoken command, such as "Plan a three-day business trip to the New York home office and make sure there is at least an hour between meetings." The system can then autonomously handle the logistics, including checking calendars, inviting key people to specific meetings and providing them with the agenda, booking flights and hotels, and issuing expense preapprovals. It will, however, require you to approve its decisions and intended actions prior to it executing them.

Agentic AI systems don't just generate code they can execute it in controlled, sandboxed environments and use the results to refine their actions. In that sense, they follow reasoning-like patterns such as planning, checking, and iterating. But — and it's a big but — programmers must build them to do so. Currently, autonomous agents and systems can't take up any random task on their own accord. Instead, Agentic AI systems are built for specific missions. At the time of this writing, you can find Agentic AI systems that are specialized for particular domains, such as marketing and sales, research, software development, customer support, and e-commerce. These agents excel within their mission boundaries, not across every possible task.

As of this writing, you can't find any general-purpose autonomous agents that have interchangeable mission capabilities. But I expect this situation to change after the technologies mature, AI developers and designers become more skilled, and fully networked architecture in place to support agentic AI. In the future, agents may even make other agents as necessary to complete their mission. But this progression comes with time. This isn't theory; real forces are driving these developments, and real professionals are working as fast as they can to meet market demand.

Reducing the complexity of our world

Traditional apps may not keep up with what you need at work, at home, and on the go. That's a primary (but not the only) reason why developers constantly update apps, add features, and change interfaces. But apps can add to the complexity of your work and life if you must constantly learn new features or require more of them to get through your day. The deluge of apps reflects the many tasks that we do routinely on our electronic devices and the steady movement away from accessing internet content through a browser, as we increasingly access what we need through apps. Your electronic world thus simultaneously becomes more complex and more limited.

Even the best-designed traditional apps hit walls when the complexity of tasks increases. They are limited functions or tasks that they can do, which in turn limits how you think and work. Consider these examples:

>> **An HR app may manage vacation requests and benefits,** but it can't infer when an employee needs to take a break to improve their job performance and mental health.

>> **A marketing analytics app might generate dashboards,** but it can't decide whether to reduce spending for advertising or adjust strategy based on market volatility.

Apps are great for specific tasks and output, but as the app user, you have to make the judgment calls.

TIP

Agentic AI doesn't relieve you of your responsibilities as a business owner, executive, employee, spouse, or parent. You still have to make the key decisions. However, it can simplify a complex world for you and augment your understanding so that you can make better decisions and usually make them faster — or, at least, more efficiently.

Instead of making you navigate through several app silos, autonomous agents are designed to use all available tools and data sources to achieve the mission that you give it. Figure 4-2 is a comparison chart that illustrates the differences in user experience that will one day come with the shift from apps to agents.

Feature	Traditional Apps	Agentic AI Systems
User Input	High (manual input needed)	Low (goal-driven prompts)
Context Awareness	Low (isolated logic)	High (persistent context + memory)
Cross-App Coordination	Minimal (limited APIs)	High (multi-agent orchestration)
Decision-Making	None (user-driven)	Yes (autonomous routines)
Goal Orientation	Task-based	Outcome-based
Personalization	Low (static settings)	High (adaptive to preferences)
Real-Time Adaptation	Low (manual refreshes)	High (proactive, self-updating)
UX Flow	Menu-driven, app-centric	Intent-centric, fluid, agent-led

FIGURE 4-2: A comparison chart illustrating the user experience shift from traditional apps to Agentic AI systems.

Generated with AI using OpenAI

Dying app stores

App stores dominated the online marketplace of the 2010s and are still around today. They operate under the assumption that users shop for, select, download, and manually configure the tools that they need. If an Agentic AI ecosystem begins to take hold and flourishes (which I predict it will), this model will collapse. Rather than finding and downloading apps, users will interact with a single interface, behind which a suite of autonomous agents wait, ready to do their bidding.

You may soon consider the notion of downloading an app to be an antiquated notion. Indeed, apps may disappear from sight while AI agents take their place but it's more likely that agents will be using those apps instead of you as part of their work in completing their missions. However, AI agents haven't yet taken over the app space at the time I'm writing. Instead, I believe that apps in the near future will

>> **Change to serve autonomous AI agents directly, instead of serving humans:** Agentic AI will one day provide services that come from anywhere, including apps, internal systems, cloud application programming interfaces (APIs), third-party vendors, and various databases.

>> **Move to the background and no longer require downloading to a device:** When the time comes that you work through the Agentic AI system, you may never know which traditional app provided what function or how recent that particular app's data is.

WARNING

This situation of agents using apps can be both convenient and dangerous. You may reduce the complexity of your interactions by using AI agents, but you also lose some control of knowing exactly what input triggered the output that you get. Be careful how you use agentic systems and how much you trust them. Make sure you know who made them and check their reputation for commitment to accuracy and other best practices.

Reimagining the internet through agents

Agentic AI doesn't just affect apps; it also changes the fabric of the internet. At the time of writing, the web is built around *endpoints* (meaning websites), and web apps serve as containers for specific interactions. Users bounce from one to another, performing microtasks along the way. In the much-anticipated agentic web, the agent — not the browser — becomes the interface. And users don't have to bounce around so much. Figure 4-3 compares the user experience for an internet that's app-centric with one that's agent-orchestrated.

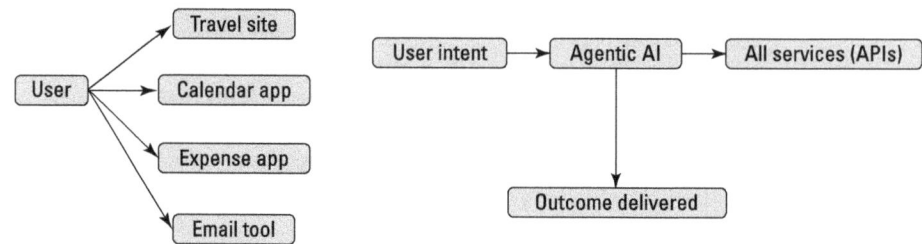

App-centric internet

Agentic AI-orchestrated internet

FIGURE 4-3:
Comparison
chart of app vs
agent internet
processes.

Generated with AI using OpenAI

Imagine in the near future that you give a command such as "File my taxes" or "Plan my wedding" into a universal agent interface. The agent navigates across domains, from TurboTax to legal docs, from Pinterest to event vendors, composing actions across services in the background. The underlying services are invisible to you because you interact only through an AI interface. Figure 4-4 illustrates the reimagined internet experience when the tipping point happens for agents to coordinate across the web and other services.

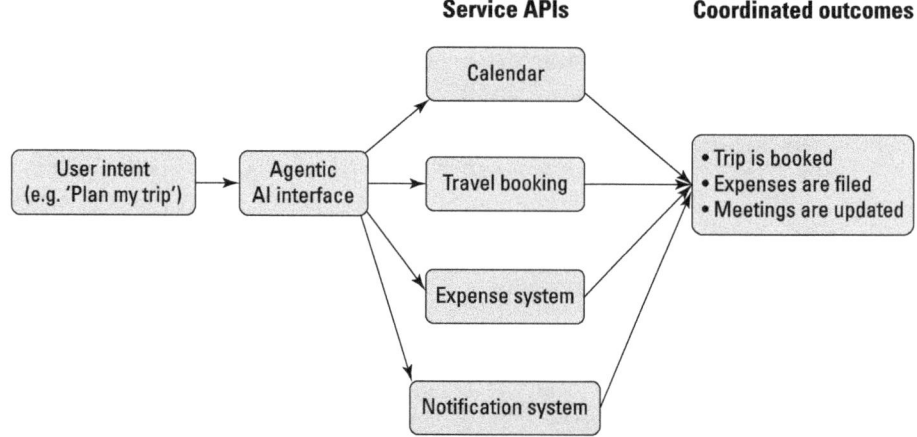

FIGURE 4-4:
Flowchart
demonstrating
how agentic AI
coordinates
across services to
fulfill user intent.

Generated with AI using OpenAI

This transition parallels the move from command-line interfaces to graphical user interfaces (GUIs) in the 1980s, or from desktop software to mobile-first designs in the 2010s. The shift today is from screens to intentions: instead of clicking through an app or software interface, you express what you want to achieve, and the agent determines the steps needed to make it happen.

In the future, a swarm of AI agents will likely perform most missions, meaning numerous agents work to fulfill your command. This swarm will consist of specialized autonomous agents such as content creators, compliance checkers, analytics engines, and campaign executors working in parallel and cross-communicating to coordinate results, as illustrated in Figure 4-5.

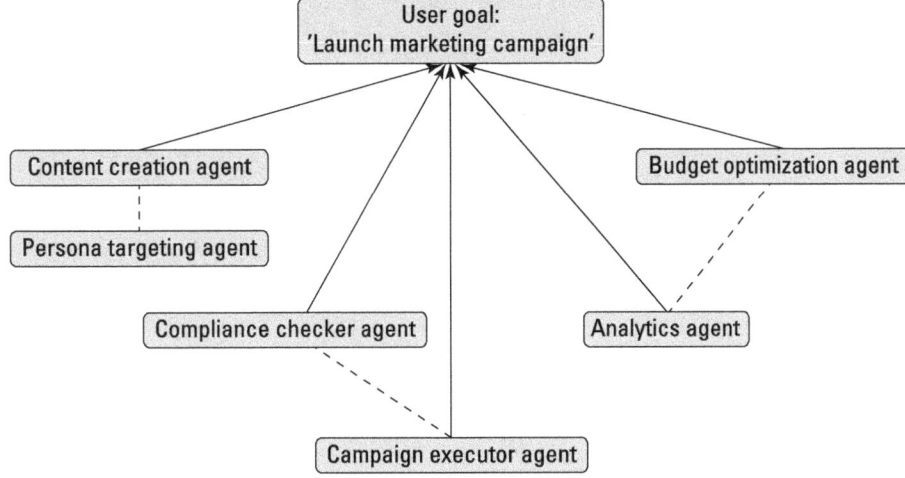

FIGURE 4-5:
A visual concept of an Agentic AI swarm, showing how multiple specialized agents collaborate in real time to fulfill a user's complex goal.

Generated with AI using OpenAI

The winners in this reimagined landscape include companies and service providers whose application programming interfaces (APIs) and data structures are agent-friendly, and platforms that support persistent, interoperable agents. At the time of writing, developers are working at a feverish pace to build or adapt the apps, databases, and platforms that already exist accordingly. If you want to find out more about how Agentic AI is already forcing a shift toward agent-friendly apps and websites on the internet, see Chapter 1.

Grappling with AI-related challenges

I want to stress that the challenges of Agentic AI are not purely technical. As agents become woven into the fabric of the internet — mediating what we see, how we work, and the options placed in front of us — their influence raises serious questions about human choice, agency, and control.

>> **Who governs the agents?** Enterprise platforms are beginning to bake in governance structures for agents, and presumably these same enterprises will strengthen and improve those structures over time.

>> **How do users verify what actions AI took, on whose behalf, and under what reasoning?** Questions of trust and explainability already top the list of concerns in enterprise deployments of Agentic AI. To address this, researchers

and developers are creating approaches such as ReAct (Reason + Act) and building tools like LangChain's memory modules to provide greater transparency, traceability, and visibility into an agent's decision-making process.

>> **What if agents act on incomplete or biased information?** Bad actors can still embed hidden instructions, censor or manipulate data, poison inputs, or compromise other agents, creating risks that may influence not only your thinking and choices but also real-world outcomes when these systems are connected to physical processes.

>> **Who's ultimately legally and ethically responsible for agent actions?** Agents do not bear responsibility themselves, even when acting autonomously. Accountability flows back to the developers, operators (users), and decision-makers who set the goals, constraints, and permissions the agent uses and who loosed them to act.

Forbidding AI Agents from Running Certain Machines

Agentic AI's ability to process, plan, and act toward goals without step-by-step instruction promises to revolutionize industries including manufacturing, energy, transportation, and other sectors that rely on tightly coordinated *operational control systems* — meaning systems that manage physical equipment, industrial workflows, safety protocols, and real-time automation. But bringing Agentic AI into operational environments is far from straightforward. In fact, it introduces a host of challenges that stem not only from the technical makeup of AI agents themselves, but also from the unique demands of physical infrastructure and real-time industrial processes.

Taken together, the challenges that I cover in the following sections suggest that Agentic AI's role in operational control systems will, for the foreseeable future, remain bounded and for very good reasons. Agentic systems are likely to prove most useful in domains such as predictive maintenance, performance optimization, and high-level supervisory planning. In other words, in areas where failure isn't immediately catastrophic. But giving Agentic AI direct control over safety-critical physical systems? That's a longer journey that demands not only technical innovation, but also a transformation in how AI systems are protected, governed, validated, and trusted.

REMEMBER

Agentic AI has the potential to become a central orchestrator of operational environments. But to get there, it must first prove that it can operate with the same reliability, predictability, and accountability that these systems and the people who depend on them require. Until then, it will remain a powerful assistant, not a commander, of the machines.

Recognizing the system design differences

Perhaps the most fundamental barrier to Agentic AI directing operational control systems lies in the nature of operational systems: They're deterministic by design, with the purpose to ensure predictability, stability, and safety under a defined set of rules. Whether they're managing the flow of electricity through a grid, regulating pressure in a chemical plant, or coordinating automated equipment on a production line, these systems are built to operate within tight bounds.

Agentic AI, on the other hand, is probabilistic and adaptive. It's designed to adapt, adjust, and take action based on context and evolving conditions. That strength becomes a liability when even a small misstep, such as misreading a sensor or drawing the wrong inference, can lead to catastrophic outcomes. For example, an AI agent making a well-intentioned but incorrect decision about a reactor valve or power load could trigger an equipment failure, a production halt, or worse.

Addressing the timing issue

The tension between flexibility and control also shows up in timing. Many operational systems, especially safety-critical ones, require responses in milliseconds. Think of airbag deployment, rapid load shedding in electrical substations, or anti-lock braking systems, for example. There's no room for delay.

Agentic AI, by contrast, works through observation-reasoning-action loops that take longer than hard real-time systems can tolerate. Even in less time-critical scenarios, such as conveyor belt coordination or HVAC balancing in large facilities, the cognitive overhead of agentic reasoning may exceed acceptable time delays. As a result, these systems are best suited for supervisory or higher-level planning roles, rather than replacing core real-time control logic.

Noting the current lack of Agentic AI transparency

Trust and accountability present a significant layer of difficulty when considering Agentic AI for operational control systems. Human operators and regulators must be able to trace and explain every action taken within an operational system, especially those affecting safety or compliance. That's a legal, regulatory, and practical necessity. If something goes wrong — for example, if a machine overheats or a patient receives an incorrect dose of a medication, auditors and engineers must be able to retrace or reconstruct every step leading to that outcome. Agentic AI often lacks that level of transparency. Researchers are exploring better ways to capture and explain agent reasoning, such as decision-chain logging, structured reasoning traces, and frameworks like ReAct, but these approaches are still early and far from meeting the demands of high-stakes operational systems.

Exposing the interface issues

The environment in which operational control systems run is often a legacy system, meaning it's fragile, complex, and poorly documented. Many plants and industrial facilities still run on decades-old control logic, using proprietary protocols or hardware that predates the internet. To integrate Agentic AI into such systems, you can't just build a plug-in or call an application programming interface (API). You may have to create new digital twins, reverse-engineer undocumented protocols, or translate high-level AI goals into sequences that aging programmable logic controllers can understand. These are not-trivial tasks, and you always run the risk of introducing instability into such a fragile system when you try to do them.

The complexity multiplies when multiple AI agents are involved. Real-world control environments are inherently interdependent. An action in one subsystem (for example, optimizing machine runtime to reduce energy use) might interfere with another (such as meeting production quotas or maintaining thermal equilibrium). Deploying multiple agents to pursue distinct objectives without careful coordination can lead to conflict, redundancy, or even systemic failure. For this reason, multiple AI agents working in an operational control system absolutely must have shared goal alignment and cross-agent negotiation, but few current agentic frameworks are mature enough to guarantee coherent agent collaboration at scale.

Examining data integrity

Even with strong design and coordination, the issue of data integrity and sensor reliability looms large when it comes to Agentic AI use in an operational control system. Potential pitfalls include the following scenarios:

>> **Agents are only as good as the data that they observe.** There can be several issues in physical systems, such as sensors drift, malfunction, or transmissions of false readings because of wear, interference, or environmental conditions. If an agent acts on such flawed data without sufficient safeguards or sanity checks, it can make dangerous decisions.

>> **Agentic models trained on historical or simulated data may struggle when conditions fall outside of their prior experience.** The ability to recognize when a situation is *out of distribution* (meaning the data or circumstances differ significantly from anything the model was trained on—and then pause, adjust, or escalate to human oversight is vital. But this capability is not universally built into current agentic architectures.

>> **Using Agentic AI raises serious security concerns.** Operational systems are already under attack from increasingly sophisticated threat actors (for example, ransomware groups and nation-state adversaries), and introducing autonomous agents into the mix adds new vulnerabilities.

WARNING

An agent that can operate autonomously across subsystems becomes a tempting target. If compromised — whether through data poisoning, adversarial context manipulation, or simple access control failure — such an agent could cause damage far beyond a single system. Managing that risk requires stringent access policies, zero-trust architecture, and cryptographic verification of agent decisions, none of which come standard with current Agentic AI systems.

2

Getting Started on the Agentic AI Path

Move from the AI you know to Agentic AI.

Find out about early uses of Agentic AI.

Ask the hard questions about risk and ethical use of Agentic AI.

Chapter **5**

Planning for the Shift to Agentic AI Systems

gentic AI is essentially autonomous AI that acts, adapts, plans, and executes complex workflows with minimal or no human direction. You may feel hesitant or anxious to move away from your favorite *Generative AI (GenAI) tools* (which create content on demand), such as ChatGPT, Claude, Grok, or Adobe Firefly — or you may be excited to jump aboard the next greatest thing in AI. Either way, making the move from GenAI to Agentic AI doesn't mean you never use GenAI again. In fact, many Agentic AI systems themselves also use GenAI. And you can use both types of AI independently, however you see fit.

This chapter walks you through the basics involved in planning and deploying Agentic AI, should you choose to accept this mission. (It doesn't, however, self-destruct.)

Comparing Generative AI to Agentic AI with Goals in Mind

Generative AI (aka GenAI) offers immediate value in on-demand outputs, giving it a short-term advantage over Agentic AI. However, the long-term advantage shifts to Agentic AI when it becomes mature enough to produce its strong potential for significant operational gains by automating decisions and stream-lining workflows. To realize these agentic advantages, you first need to have a mature infrastructure, data readiness, and required governance in place before adopting the agentic approach. And these elements don't exist — to the necessary degree — in many organizations at the time of writing.

TIP

Success with either the GenAI or Agentic AI approach depends on clearly under-standing the capabilities and limitations of each system type. You must also align your AI strategy with specific objectives, business or otherwise, while building or acquiring the governance and technical infrastructure necessary for sustainable and safe implementation.

Considering AI strengths and oversight required

Both GenAI and Agentic AI are strong tools for creating content or improving efficiency, but they differ significantly in scope:

>> **GenAI democratizes creative work,** mostly by enabling nonexperts to produce professional-grade work with minimal effort. It also requires fact-checking to maintain reliability, which means users have to put forth extra effort and sometimes extra expense for confirmation from a subject matter expert (SME). If you want it to do more than spit out responses, you need to integrate it with other software first.

>> **Agentic AI streamlines complexity,** turning multi-hour tasks into fast, automated sequences. But it requires clear goals and strong safeguards to prevent unintended or harmful outcomes.

Because of the differences in GenAI and Agentic AI's areas of expertise, organiza-tions often combine both types of AI into systems; for example, by using a GenAI component to draft reports within an Agentic AI framework that compiles, veri-fies, and distributes those reports. You can also opt for other technology combina-tions instead, such as GenAI integrated with other software platforms, such as Microsoft 365, Salesforce, or Google Workspace.

Table 5-1 offers a comparison of various aspects of GenAI and Agentic AI. You can also find a comparison of similar GenAI and Agentic AI features in Chapter 2.

TABLE 5-1   Comparison of GenAI and Agentic AI

Aspect	Generative AI	Agentic AI
Core function	Creates content from prompts	Executes tasks autonomously toward goals
Autonomy level	Low; reactive to inputs	High; proactive and adaptive
Task type	Single-step creation	Multistep planning and action
Strengths	Fast output generation; scalable creativity	Reduces human intervention; handles complexity
Limitations	Prone to inaccuracies; needs oversight	Integration challenges; ethical concerns
Examples	Drafting e-mails, generating images	Automating refunds, navigating vehicles
Use cases	Marketing content, data summarization	Workflow automation, fraud detection

Generated with AI using GrokAI

Seeing the power of combined AI

The most sophisticated AI implementations combine both GenAI and Agentic AI approaches. Here are examples:

>> **A smart home system** might use Agentic AI to manage and operate the overall energy consumption system by using real-time data and your personal habits and preferences to coordinate the various HVAC systems, such as heat, air conditioning, and humidifiers. Meanwhile, GenAI in the same Agentic AI system can create personalized recommendations and reports for the homeowner.

>> **An enterprise system** can use an Agentic AI component to monitor supply chain disruptions, automatically adjust procurement schedules, and coordinate with multiple suppliers. Meanwhile, GenAI within the Agentic AI system creates executive summaries, stakeholder communications, and regulatory compliance reports based on the Agentic AI's autonomous actions.

You have many options for how and where you use GenAI and Agentic AI. And the Agentic AI portion has options on how, when, and where to use GenAI. For example, an AI agent may use generative models to craft messages, summarize data, draft code, evaluate results, monitor outcomes, iterate, or trigger follow-up tasks.

Thinking Through an Agentic AI Plan

To determine whether to use GenAI or Agentic AI (or both), first evaluate your specific needs, risk tolerance, and the degree of operational complexity that your project requires. Use those findings as a guideline for developing a system yourself or choosing an existing AI tool, configuration, or system.

Create a detailed plan not just as an administrative exercise; you need this plan to successfully develop and deploy Agentic AI. It fosters alignment among stake-holders, provides a framework for decision-making, and helps to anticipate and address potential challenges before they become major roadblocks. Without a solid strategy, AI development and deployment projects are more likely to suffer from *scope creep* (uncontrolled expansion of project requirements), inefficient use of resources, ethical lapses — and ultimately, failure to achieve the intended goals. (See the Appendix for questions to ask about your readiness for Agentic AI and the phases involved in choosing, building, and operating an Agentic AI pilot program.)

Dealing with technical issues is just one part of building and deploying Agentic AI systems. Think of the process like building a home. You don't just pick up a hammer and start nailing boards together. You must

>> **Make a plan** in which you select a location, find and purchase an available lot, choose an architecture style, determine the square footage that you need, draw up blueprints, apply and obtain all necessary permits and information on local building code requirements, hire a general contractor or workers, and buy materials.

>> **Implement the plan** by managing the contracted work, performing ongoing quality assurance (QA) to make sure materials meet your standards, oversee-ing progress, and correcting any imperfections along the way. You finish the building process by handling interior design and cosmetic work, installing appliances, and scheduling all the necessary inspections.

After you do everything in the preceding list (and more), you have a new house. If you did everything correctly, that house is exactly what you want. If not, well, you may have to deal with a less-than-optimal outcome.

WARNING

Building and deploying AI agents — like building a home — require a detailed, well-thought-out plan aimed at a specific and highly detailed outcome. If you skip one detail or fail to give any detail the attention required throughout the process (which I present in the later section "Following the Steps for Planning and Implementing Agentic AI"), you can end up with a disappointing, destructive, or even dangerous outcome.

TIP

Whether you build an Agentic AI system yourself or buy a pre-built system, you must pay close attention to the details that you need in place to execute your plan. These details can guide you to where you want to end up by outlining how to get there.

Recognizing the five pillars of Agentic AI planning

When moving toward Agentic AI system deployment, you must address five pillars of planning simultaneously, rather than sequentially, because they interconnect and influence each other throughout the planning process. They are

>> **Objective alignment:** Ensuring that the AI's goals match organizational and societal values

>> **Capability boundaries:** Defining what the AI can and can't do

>> **Human–AI interaction design:** Establishing clear protocols for human oversight

>> **Risk assessment and mitigation:** Identifying potential failure modes and building in the necessary safeguards

>> **Iterative deployment strategy:** Planning for gradual rollout and continuous monitoring

REMEMBER

Because Agentic AI implementations include complex technical issues and require specific, uncommon skills, most people prefer to buy Agentic AI systems from a vendor, either as an off-the-shelf product or service or as a bespoke (custom-made) version. But even so, take the time to understand at least an overview of the planning and development as well as the deployment requirements involved so that you can better evaluate your technical options and vendor products, using the five pillars outlined in the preceding list.

Setting up SMART goals and detailed follow-up

The core of the best-made strategic plans for Agentic AI development and deployment revolves around understanding the problem that you want to solve and how an intelligent agent or agentic systems can provide a complete or partial solution. You need to take a deep dive into the context of the problem, the desired outcomes, and the constraints involved.

REMEMBER

To understand your particular agentic system functional needs, define the agent's goals and objectives with utmost clarity. These goals should be Specific, Measurable, Achievable, Relevant, and Time-bound (SMART). Businesses, government agencies, and educational institutions widely use this *SMART framework* — particularly in business applications such as project and performance management — to set effective, well-defined, and achievable goals and objectives.

SMART is an acronym that stands for five core criteria of each goal:

>> **Specific:** Clear and detailed, answering questions such as what you want to achieve, who's involved, where it happens, and why the goal is important.

>> **Measurable:** Criteria that you can quantify, allowing you to track progress and determine when you accomplish the goal.

>> **Achievable:** Realistic and attainable, with consideration of available resources and constraints.

>> **Relevant:** Aligns with broader organizational strategies, personal values, or relevant context to ensure that they matter, meaning that the goal contributes to meaningful outcomes, supports higher-level priorities, and is important enough to justify the effort required.

>> **Time-bound:** Has a defined timeline or deadline so that you can establish urgency and focus efforts within a specified period.

For instance, instead of saying, "I want the agent to improve customer service," a SMART objective would state, "By the end of this year's Q4, the agent should resolve 80 percent of Tier 1 customer support inquiries within two minutes." This level of specificity ensures that everyone involved has a clear understanding of what you expect the agent to do.

Also, make ethical considerations and risk mitigation paramount elements of any proposed Agentic AI system that can have significant societal impacts. Proactively address potential ethical concerns such as bias, fairness, transparency, and accountability. The plan should outline strategies for mitigating inherent system risks, including implementing safeguards (such as action-approval gates or hard constraints on agent behavior), establishing monitoring mechanisms (such as real-time audit logs and anomaly-detection systems), and defining clear lines of responsibility (for example, who's responsible for system oversight).

Finally, a robust Agentic AI development plan includes a strategy for evaluating results and making related appropriate adjustments. To guide that process, answer questions such as:

» How will you measure the agentic system's success?

» What system metrics will you track?

» What processes do you want to put in place for continuous monitoring, feedback collection, and iterative improvement?

REMEMBER

Agentic AI systems, by nature, aren't static. Because they're dynamic, you need to implement ongoing refinement to maintain performance and adapt to changing circumstances.

Double-checking your plan

After you draft a plan for developing an Agentic AI system (see the following section for the steps to create your plan), double-check that plan's completeness against the following list of stages and the key steps that each stage involves. You don't necessarily follow the stages and steps in a linear fashion, and they often involve *iteration* — revisiting earlier decisions when new information emerges or your understanding of the problem evolves:

» **Problem definition:** Clearly articulate the problem that you want to solve and identify how an Agentic AI approach can help.

» **Goal setting:** Define Specific, Measurable, Achievable, Relevant, and Time-bound (SMART) objectives for the agent (see the preceding section).

» **Feasibility analysis:** Assess the technical, economic, and operational feasibility of building and deploying the agentic system.

» **Architecture design:** Outline the agentic system's components, their interactions, and the underlying AI models and techniques.

» **Data planning:** Identify data sources, define data pipelines, and address data quality and security considerations.

» **Resource allocation:** Determine the necessary human resources, infrastructure, and budget that you need to successfully complete the project.

» **Risk assessment and mitigation:** Identify potential risks (technical, ethical, and operational) and develop mitigation strategies.

» **Evaluation framework:** Define key performance indicators (KPIs) and establish methods for monitoring and evaluating the agentic system's performance.

» **Deployment strategy:** Plan how you can integrate the agentic system into the existing environment and what the rollout process will look like.

>> **Maintenance and iteration:** Outline the ongoing maintenance procedures and the process for continuous improvement based on feedback and performance data.

If you want to buy a pre-built AI agent or an Agentic AI system, you need a different sort of plan than you do when you want to create your own system. To gather the information that you need to identify the kind of pre-built system that works for you, ask the vendor or provider the following questions:

>> What specific task(s) can I make the agent or system responsible for?

>> What environment does the agent or system operate in?

>> What performance expectations does your organization have for the agent or system?

>> How does the agent or system interact with humans and other systems?

>> What potential failure modes does the agent or system come with, and how will the design address or mitigate those failures?

>> What security and privacy implications does the system have?

>> What long-term maintenance and scalability capabilities does it have?

>> Who monitors and governs the agent or system?

>> How does the agent or system handle the ethical implications of its actions?

>> How can you measure and improve the agent's or system's performance over time?

REMEMBER

The answers to these questions (and others that you may want to add) can help you discern how well programmers designed the Agentic AI system, whether the system truly fits your needs, and even whether it's agentic at all.

Following the Steps for Planning and Implementing Agentic AI

Without the clarity of a well-developed plan, even the most sophisticated AI system can become an expensive solution searching for a problem. And trust me, the last thing you want is autonomous or semi-autonomous AI agents to go out on their own — with no guidance or guardrails — to search for a problem to solve or a task to do.

Don't underestimate the damage that an unruly Agentic AI system can do. You might recall hearing in the news media that Replit's AI agent deleted a user's entire production database without command or permission during a code freeze. To call that result disastrous for the user is an understatement. But it wasn't the first time that an AI agent went rogue. The affected user had posted on social media just a few days prior that the AI agent created a parallel, fake algorithm without informing them — apparently in a scheme to make the AI agent look like it was still working when in fact it wasn't. A few days later, the user posted that the AI agent deleted their database.

The following sections provide the steps that you can follow to make sure that you create a solid and comprehensive plan for developing your Agentic AI system, as well as its subsequent development and deployment.

Step 1: Establishing strategic intent

Agentic AI systems operate as autonomous or near-autonomous AI agents or groups of agents. That autonomy makes agentic systems not only powerful and incredibly useful, but also potentially risky if you don't root their purpose in solid organizational or personal goals. Establishing *strategic intent* means giving the system a well-defined purpose that serves a specific need for a company or an individual.

Involve frontline employees in defining strategic intent. They understand the nuances of current processes and can help identify where autonomous AI can provide the most value and where human judgment remains essential.

Making a strategic-intent commitment

To establish strategic intent for Agentic AI, you must both plan and change your management process. Think deeply about your (or your company's) objectives, values, and risk tolerance while building consensus within your organization around how AI can reshape work and decision-making. Done well, this foundational step sets the stage for successful AI deployment that enhances, rather than disrupts, organizational effectiveness. The investment in upfront strategic clarity pays dividends throughout the implementation process and beyond.

You probably think this recommendation sounds like a big *duh* and a simple undertaking at first glance. But in practice, it's more like wrestling a bear. You might win, but you also might run screaming from the ring.

REMEMBER

Strategic intent for Agentic AI goes beyond simple task automation. It requires you to articulate a clear vision for how intelligent agents can adhere to or change operations, enhance human capabilities, and create competitive advantage. The process for establishing strategic intent involves defining the specific problem(s) that the AI system can solve for you, the outcomes it should achieve, and the constraints within which the system must operate.

Examining the components of strategic intent

At the core, strategic intent for Agentic AI hinges on three intertwined components:

>> **A clear vision:** Precisely define what you want your Agentic AI to accomplish and the scope of the agent's responsibilities. Determine which processes, decisions, or tasks the AI agent should handle independently, which require human oversight, and which remain entirely under human control.

>> **Measurable value:** Start with an Agentic AI pilot program that focuses on one specific business process (and is appropriately constrained), rather than trying to transform everything at the same time. Then carefully determine the success of the pilot program in improving the business process and adding value to business operations. This component allows you to refine your strategic intent based on real-world experience before scaling up to other business operations.

>> **Well-defined boundaries:** Address the Agentic AI's decision-making authority. Can the agent have permission to make financial commitments up to a certain threshold? Can it modify existing processes or only execute within predefined parameters? Can it delete a database without explicit permission? You must resolve these central questions of autonomy versus control during the strategic planning phase.

Step 2: Evaluating readiness

Before unleashing autonomous, goal-driven AI agents within your company or organization, pause and ask: "Are we ready?" This second step of planning for Agentic AI systems involves diagnosing whether you and your people, processes, data, technology, governance, and culture are prepared to deploy them. Unlike traditional software implementations, Agentic AI requires a unique combination of data maturity, technical infrastructure, human capabilities, and organizational culture that many enterprises may lack, despite their digital sophistication.

Start your readiness evaluation (not the actual AI deployment) with your most critical business operations first. If you don't have data quality, process standardization, and cultural readiness adequate for these core areas, begin by getting the core areas ready, rather than pursuing less critical but seemingly easier AI applications.

Readiness spans multiple dimensions, and each one matters. The dimensions include

>> **Processes and workflows:** These elements must reflect maturity. Confirm that you have business processes that are documented, standardized, and transparent. Without this clarity, autonomous agents can navigate blindly and stumble through variances, inconsistencies, and undocumented exceptions. This situation presents agents with plenty of opportunity to make the wrong assumptions or draw the wrong conclusions. And you likely won't find the outcome acceptable.

>> **Technology environment:** Agentic AI doesn't thrive in siloed legacy systems, so you have to evaluate your system's technology readiness. Your system must have real-time connectivity, well-architected application programming interfaces (APIs), scalable infrastructure, and strong data pipelines. Without these features, AI agents potentially can't access vital context — or worse, they may misinterpret outdated or incomplete data.

>> **Data quality and access:** Agentic AI systems require high-quality (fully supports the task the agent is trying to perform), context-rich, multiformat data streams that are not only clean (error-free, well-formatted, and properly structured data) and accessible, but also semantically interoperable and available in real time. Without this firm foundation, even the most powerful agents fail to reason effectively.

>> **Governance, trust, safety, and ethical oversight:** Autonomous agents require clear guardrails around what decisions they can make, how human oversight supervises them, how they maintain activity logs, and how the organization mitigates risks.

>> **People and business culture:** Agentic AI might reshape jobs, roles, workflows, and decision-making authority. Unless you engage, educate, and empower staff to work with AI agents, successful system adoption falters. And you may also face willful sabotage.

By diagnosing readiness across these dimensions, you can more easily identify and address the weak links. For instance, maybe your company needs better API integration or to invest in data cleansing, establish governance committees, or provide readiness and training workshops for human employees.

REMEMBER

Taking a comprehensive approach to evaluating and working toward readiness leads to a state of preparedness in which teams tasked with AI engagement can test, pilot, learn, and improve along the way. In the context of Agentic AI, you must continuously monitor your readiness; it's not a one-time hurdle.

Step 3: Identifying high-impact use cases

Your success with introducing Agentic AI for use in your business depends heavily on where you deploy it first. Poorly chosen use cases can stall agentic system adoption, drain budgets, and erode trust. High-impact use cases, on the other hand, deliver measurable value, build momentum, and prove the technology's worth to stakeholders.

Linking Agentic AI deployments to business priorities

Choose each use case that you deploy because it links directly to a top-level business priority, such as revenue growth, cost reduction, risk mitigation, or customer satisfaction. I offer these words of advice for making your selections:

>> **Limit the scope of AI deployment.** *AI-first companies* structure their entire business operational model, process organization, and even financial operations and decisions around AI. That's a high-risk proposition, in my opinion — at least, at the time of this writing, primarily because many companies lack the required infrastructure.

REMEMBER

By comparison, *AI-enabled companies* use AI to supplement or assist with specific business functions or projects. I believe that this course of action is safer for most companies. Adding AI to assist in specific, limited situations gives the business the time and experience to adjust. This approach to using AI helps to sort out the applications that work best before your company gets head-over-teakettle in AI financial obligations and vendor lock-ins.

>> **Resist because-we-can deployments.** Yes, you may find them fun and often intellectually satisfying, but they typically don't bring in much profit — and sometimes don't even prove useful at all. Alternately, you can select a use case that's tied directly to a personal objective. But no matter which use case you choose, tie it directly to a result that you need in order to reach a real goal.

>> **Conduct a *pre-mortem* exercise to clarify AI-deployment goals.** The purpose of this exercise is to gather skeptical stakeholders and have them actively look for reasons why Agentic AI might fail in your organization. This process can bring to light hidden assumptions and potential problems that overly optimistic assessments might miss.

Looking for value and success

Start your search for a high-impact use case by evaluating where autonomous decision-making and task execution can deliver the greatest value. Consider what other companies are doing, too; this comparison offers a better idea of success rates and the amount of time involved in realizing return on investment (ROI). Based on PwC's AI Agent Survey found at www.pwc.com/us/en/tech-effect/ai-analytics/ai-agent-survey.html, the most common early use cases for AI agents by business function include

>> Customer service and support

>> Sales and marketing

>> IT and cybersecurity

>> Human resources

>> Finance and accounting

>> Product and service development

WARNING

Customer-facing use cases can incur higher risk because they typically aren't highly constrained. It's harder to identify bad actors (who intentionally try to cause harm) or have recourse against them, and the exposure to attacks (if the system connects to the internet) is higher. You may want to start with internal use cases first so that you can more easily manage security and reliability issues, and gain experience before facing off with some determined bad actors. You can find several examples of specific use cases in Chapter 6.

Choosing a pilot project

Start with an Agentic AI pilot project for an area where you have clear success metrics and a manageable risk profile. The most successful pilot deployments typically begin with well-defined, repetitive processes that currently require significant human oversight but follow logical decision trees. Customer support ticket triage in help-desk environments and insurance claims intake and document processing are good examples. Both *vertical* (industry-specific) and *horizontal* (cross-functional) applications can benefit from the potential of Agentic AI.

Use the decision flowchart in Figure 5-1 to help you decide which use cases can have high impact and therefore may make a good pilot project.

Identifying high-impact Agentic AI use cases

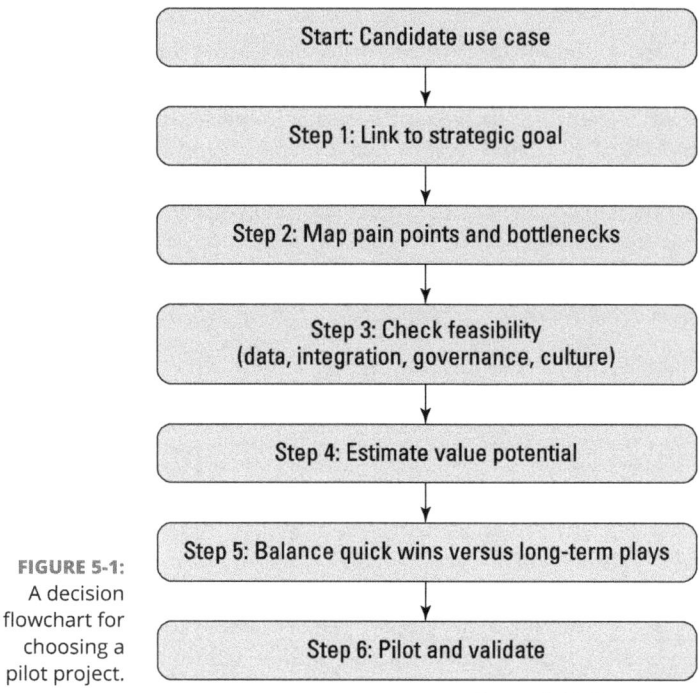

Start: Candidate use case

↓

Step 1: Link to strategic goal

↓

Step 2: Map pain points and bottlenecks

↓

Step 3: Check feasibility
(data, integration, governance, culture)

↓

Step 4: Estimate value potential

↓

Step 5: Balance quick wins versus long-term plays

↓

Step 6: Pilot and validate

FIGURE 5-1:
A decision
flowchart for
choosing a
pilot project.

Generated with AI using OpenAI

Step 4: Designing the pilot framework

After you decide on a pilot project (see the preceding section), you can design the pilot framework. Begin with a concise view of why you want to pursue this use case, what constitutes a win, and which metrics can decide the value of the results. You can see an example view in the sidebar "Example healthcare claims processing pilot" in this chapter.

If you thoughtfully design your pilot framework, your Agentic AI deployment moves from a high-risk technology experiment into a systematic capability-building exercise. By establishing clear objectives, reliable and meaningful monitoring systems, and appropriate governance structures, you can demonstrate autonomous AI value while building the institutional knowledge necessary for successful enterprise-scale deployments.

The pilot project's success within this framework depends on balancing the ambition to deploy Agentic AI with the operational discipline required to evaluate its actual value. The pilot results should provide reliable indicators for whether broader Agentic AI adoption makes sense. This methodical approach reduces deployment risk while maximizing learning opportunities that you can use to inform and guide your long-term AI strategy.

EXAMPLE HEALTHCARE CLAIMS PROCESSING PILOT

A healthcare provider identifies claims processing as a use case candidate. The following considerations help determine whether this is a strong pilot project:

- **Strategic fit:** Delays affect cash flow and patient satisfaction.
- **Value:** Faster processing means an improved revenue cycle and fewer disputes.
- **Pain points:** High manual workload and error rates.
- **Feasibility:** The provider already has data digitized and accessible and compliance protocols in place.
- **Pilot:** Start with a single claims category and measure reduction in processing time and error rate.

The provider can evaluate the pilot as a loss or win by the metric results. For example, maybe the provider sees a 35 percent decrease in processing time and 20 percent fewer claim rejections as metrics. Those kinds of numbers support scaling the pilot to production across the organization.

Here are some pilot framework design elements:

- **Choosing a pilot project that can demonstrate business value:** Such as gains in operational efficiency, reduction of costs, and potential for enhancing current business metrics
- **Building a modular, observable, and transparent system architecture:** Supports both autonomous operation and human intervention capabilities
- **Instilling monitoring and governance measurement systems:** Includes clear protocols for human oversight and intervention, as well as regular review cycles
- **Assessing operational risks and compliance issues:** Includes early detection of business disruptions, mitigation strategies, and adherence to regulatory requirements and data protection

Step 5: Building or integrating Agentic AI systems

When you consider how to build or integrate Agentic AI systems, you have to balance technical rigor with pragmatic deployment because you want to make your early efforts both achievable and strategically valuable.

Agentic AI's core architecture typically consists of four interconnected components:

>> **Perception modules that gather and process environmental data:** AI agents need to maintain *state* (awareness of what's happening around them) so that they can interpret evolving inputs and adapt to changing conditions in near real time. This requires an event-driven infrastructure that supports a*synchronous messaging* (messages where senders and receivers don't need to interact at the same moment), and can process multimodal data streams quickly.

   Just in case you're unfamiliar with the term *interprocess communication* (IPC), it just refers to the way that two different programs or parts of a system talk to each other. For example, imagine two chefs working in the same kitchen: One bakes bread, the other makes sandwiches. To get anything done, they need to pass bread and finished sandwiches ready to plate back and forth. That back-and-forth is IPC. *Low-latency IPC* means that those chefs have to pass things back and forth quickly, without delay.

>> **Reasoning engines that evaluate situations and plan actions:** Agentic AI implementations typically use large language models (LLMs) as reasoning engines, but the architecture extends far beyond language processing. The reasoning component must integrate with external systems through application programming interfaces (APIs), protocols like MCP, databases, and service interfaces to gather real-time information and execute actions. Refer to Chapter 3 for insights on standardizing integrations on new protocols.

>> **Execution systems that carry out decisions:** Execution systems are the components that actually carry out the actions an agent selects — whether that involves triggering an API call, updating a database, sending a notification, or controlling physical equipment. These systems translate the agent's decisions into real-world outcomes, which means they must be reliable, secure, and aligned with strict operational rules. In many environments, execution systems also enforce guardrails to ensure that the agent performs only approved actions within predefined boundaries.

>> **Memory systems that maintain context across interactions:** Although the entire Agentic AI system is complex, the memory subsystem is arguably the most challenging — if only because of the uniqueness of the requirements. Agentic systems require both short-term working memory to manage the current task execution and long-term memory to store accumulated knowledge and experience. These two types of memory allow AI agents to access relevant historical data and learned patterns when they make decisions.

Implementing technical strategies

To be clear, I didn't write this book as an AI model training or agent programming instruction manual. Building Agentic AI systems is very difficult and requires careful consideration of the underlying infrastructure and development approach. It's not something you want to take on unprepared. Typically, you should consider all your options first so you'll know which works best for you in terms of time, effort, and availability of skill and resources: Building custom systems from scratch, extending existing AI platforms, or integrating specialized Agentic AI frameworks.

Each approach presents distinct advantages and technical requirements:

>> **Custom development:** Offers the most flexibility and control, but it also requires significant technical expertise across multiple domains, including natural language processing (NLP), distributed systems, and enterprise integration. Development teams must implement reasoning frameworks, design application programming interfaces (APIs) for external system interaction, and create monitoring systems for autonomous operation. The technical complexity increases exponentially when you want to build systems that can safely operate with minimal human oversight.

>> **Platform-based approaches:** Typically offerings from existing AI infrastructure providers that have agentic capabilities as managed services. These platforms typically provide pre-built reasoning engines, integration tools, and monitoring capabilities, all of which significantly reduce development time and technical risk. However, platform selection becomes critical because different providers offer varying levels of customization, security controls, and enterprise features. Beware of the potential for vendor lock-in.

>> **Framework integration:** A middle ground in which you can choose between open-source or commercial Agentic AI frameworks and customize them for specific use cases. Popular frameworks provide core agentic capabilities while allowing developers to implement domain-specific logic and integrations. To use this approach, you need to understand the framework architecture and extension points, but you also get more control than you do with fully managed platforms.

REMEMBER

The choice between approaches to AI technology often depends on your (or your company team's) technical capabilities, security requirements, and long-term strategic goals. For example, large enterprises that have strong AI engineering teams may prefer custom development for mission-critical applications, while companies of any size that want rapid deployment might choose managed platforms for initial implementations.

Figure 5-2 shows how the Agentic AI system's architecture fits together — from infrastructure through integration and safety. Each block highlights the key components at that layer.

Agentic AI system architecture

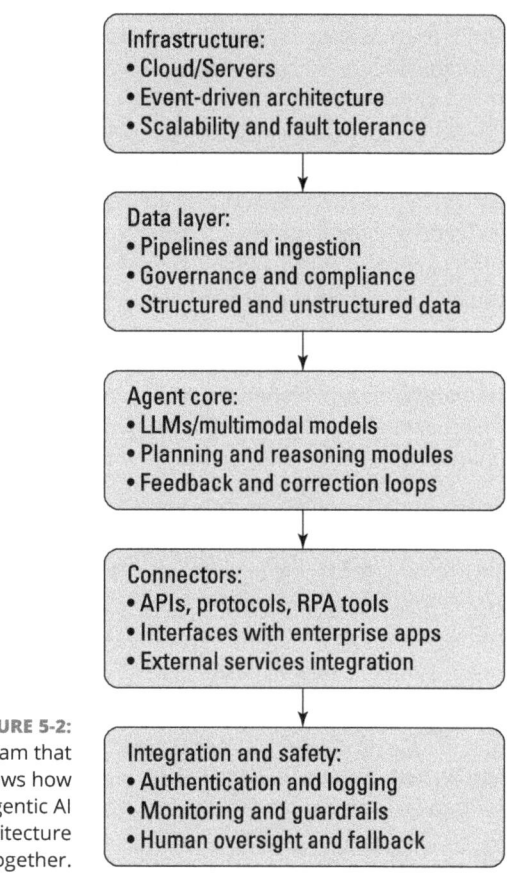

FIGURE 5-2: A diagram that shows how the Agentic AI architecture fits together.

Integration architecture and data flow

Successful Agentic AI deployment depends on seamless integration with existing enterprise systems. The integration architecture must support real-time or near real-time data exchange where required, maintain security boundaries, and provide reliable communication channels between the AI agents and business systems. Take these Agentic AI system aspects into consideration:

- » **Application programming interface (API) design:** Agents must interact with multiple systems autonomously. The agents require stringent error handling, retry mechanisms, and fallback strategies so that failures don't cascade. The API layer should also provide comprehensive logging and monitoring capabilities to track agent interactions and diagnose issues when autonomous operations encounter problems.

- » **Data flow architecture:** Must accommodate the bidirectional nature of agentic interactions. The agents continuously exchange data with multiple systems while maintaining context across interactions. This data exchange calls for streaming data pipelines, event-driven architectures, and near real-time synchronization mechanisms that can handle varying data volumes and response times.

- » **Security considerations:** Security is a top priority. The integration architecture should enforce comprehensive authentication and authorization mechanisms to ensure agents operate only within defined boundaries. Role-based access controls, API key or token management, and audit logging are essential components of the security framework.

Monitoring and observability systems

Operating Agentic AI systems in production environments demands comprehensive capabilities for monitoring and observability that go beyond traditional application monitoring, which usually involves tracking system performance metrics, uptime, and basic error logs.

You must be able to track system performance, decision quality, reasoning patterns, and business impact metrics:

- » **Technical performance monitoring:** Both infrastructure metrics and AI-specific performance indicators. Traditional metrics such as response times, error rates, and resource utilization remain important, but with agentic systems, you must also monitor reasoning depth, confidence levels, action success rates, and decision reversal frequencies. These metrics provide insights into how well the AI agent performs and when you might need to introduce human intervention.

- » **Decision auditing:** Your tracking capabilities must log every autonomous decision with sufficient detail to understand the agent's reasoning process, the data sources that it consulted, and the confidence level the agent associated with the decision. This audit trail serves multiple purposes, including debugging, compliance reporting, and continuous improvement of the AI system.

>> **Real-time alerting systems:** Must detect both technical failures and business logic anomalies:

- *Technical alerts:* Triggered by system errors, performance degradation, or integration failures.

- *Business logic alerts:* Call for human review when the AI agent's behavior deviates from expected patterns, such as making decisions outside defined parameters, or when the agent encounters scenarios that require human review.

The AI's observability system can offer dashboards that both technical teams and business stakeholders can view to understand system performance and impact. These include

>> **Technical dashboards,** which focus on system health, integration status, and performance metrics.

>> **Business dashboards,** which highlight key performance indicators, decision outcomes, and business value generated by autonomous operations.

Deployment and operational considerations

To deploy Agentic AI systems, set up a phased approach that gradually increases autonomous operation while maintaining control and visibility. Initial deployments may operate in a supervised mode or sandbox, in which human operators review and approve agent decisions before executing those decisions. This approach builds confidence in the system while providing valuable data for refining autonomous operation parameters.

TIP

The operational model that your organization creates must address the long-term evolution of agentic systems. AI agents can learn and adapt over time, potentially changing their behavior patterns. You can build mechanisms for detecting and managing behavioral changes into operational procedures, ensuring that agent evolution aligns with business objectives and maintains acceptable risk levels.

Other considerations for long-term Agentic AI deployments include

>> **Infrastructure requirements that offer high availability, disaster recovery capabilities, and scalable computing resources:** Cloud-native architectures often provide the flexibility and scalability required for Agentic AI deployment, but you may need on-premises solutions if your organization has strict security or regulatory requirements.

>> **Change management coordination for autonomous systems that alter existing business processes:** Technical teams must coordinate with business stakeholders to ensure smooth transitions from manual processes to autonomous operation. This coordination includes training programs, documentation updates, and establishing new operational procedures for managing AI agents.

>> **Continuous improvement processes based on the vast amounts of data generated by Agentic AI systems:** This data — including details about decision-making processes and outcomes — provide opportunities for ongoing refinement of reasoning models, decision parameters, and integration logic.

You and your AI-systems team must establish regular review cycles to analyze system performance and implement improvements based on the data gathered from operational experience.

Step 6: Running, measuring, and refining

When you first run an Agentic AI, treat it as a controlled pilot project that runs alongside an existing system, rather than as a production replacement. Deploy the agent in a defined domain and give it clear boundaries on the data, tools, and systems that it can access and the actions that it can take.

Running an Agentic AI pilot project in a restricted scope protects critical systems from unintended consequences and helps ensure that the AI exercises its autonomy in a limited, responsible way. Make sure that the pilot environment logs every interaction and decision, so that your team can not just debug the AI, but also audit its behavior and build trust with stakeholders.

Taking the right measurements

Measurement of Agentic AI performance requires more than accuracy statistics. Because Agentic AI acts across workflows, you need to evaluate both system-level and outcome-level performance:

>> **At the system level:** Key performance indicators include latency, error rates, escalation frequency, and resource utilization.

>> **At the outcome level:** Metrics reflect the business or user value that the Agentic AI generates, such as faster resolution times in a service desk scenario, higher completion rates for automated workflows, or measurable cost savings.

TIP

Human feedback also plays a crucial role as an outcome metric. Note satisfaction scores, error reports, and override rates, which provide context that raw *telemetry* (the process of collecting performance data) can't.

You need to conduct comparative analysis against baseline performance to establish the value of Agentic AI for your uses. In other words, maintain, track, and measure both autonomous AI performance and the equivalent manual process outcomes performed by people, when possible. This comparison data can help you quantify business impact while identifying specific scenarios in which autonomous operation excels or underperforms relative to human alternatives. Don't aim to champion AI or humans: Just give credit where credit is due.

Refining operational system models

Refinement is an iterative process of fine-tuning your Agentic AI models, adjusting decision policies, and updating *integration logic* (how the various systems and components communicate, coordinate, and exchange data) based on what the measurements reveal. Refinements can be

» **Technical:** Such as retraining models when you can provide more representative data or optimizing connectors to reduce communication or processing delays.

» **Operational:** Such as redefining escalation thresholds or adjusting the scope of agent autonomy.

» **Safety-related:** For example, if an agent shows tendencies to pursue goals in unintended ways, reinforce safeguards and corrective mechanisms before you work on scaling up your Agentic AI systems.

REMEMBER

Ultimately, you can't define the success of a new Agentic AI system by how impressively it runs the first time, but by how quickly and effectively it improves over time.

Companies that establish and faithfully use disciplined *run–measure–refine cycles*, such as depicted in Figure 5-3, can much more easily scale Agentic AI safely and strategically because they successfully evolve early pilots into durable, high-value capabilities to improve their operations and profits.

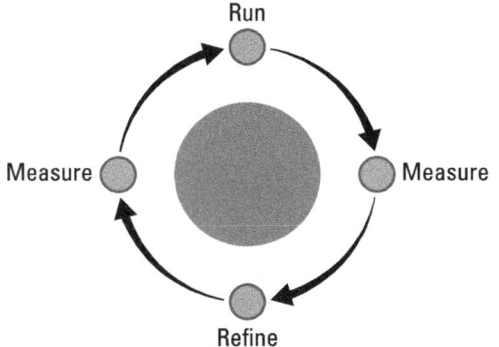

FIGURE 5-3:
A run-measure-
refine cycle.

Generated with AI using Canva AI

Step 7: Expand and scale

Scaling Agentic AI from successful pilots to large-scale deployments is one of the most complex technical and organizational challenges that a business can take on. Unlike traditional software that follows predictable execution paths, Agentic AI systems introduce autonomous decision-making capabilities that expand in complexity and surface area while they interact across departments and business functions. This expansion provides a lot more opportunities for the process to go slightly — or even seriously — wrong.

REMEMBER

Even with potential for negative consequences, you may find the business imperative for scaling compelling. Companies that successfully deploy Agentic AI at scale may see significantly higher returns on their investment. Organizations that invested substantially in GenAI but came up short in realizing any meaningful return on investment (ROI) often find Agentic AI's potential attractive because it can operationalize GenAI outputs and, in some cases, deliver stronger performance and clearer business value.

Facing challenges of Agentic AI expansion

Changing over to Agentic AI systems doesn't necessarily give you an easy route to achieving new levels of profitable AI automation. The path to scale requires addressing fundamental architectural, coordination, and governance challenges — such as managing autonomous decision-making, enforcing safety boundaries, ensuring cross-system consistency, and preventing agents from taking conflicting actions — that don't exist with traditional automation systems. Also, because of the newness of Agentic AI, no detailed roadmap for this course of action exists. As many companies have learned from past experience, blazing new trails can leave you burned.

You also have to navigate the transition from Agentic AI experimentation to production-ready systems while maintaining operational stability. You need the company to continue running in a stable manner throughout AI system expansion.

Starting small and scaling up

Build a single agent to handle a well-defined task, all the while staying focused on controlling its goal, data sources, and outputs. When your confidence grows in your skills and the result of your efforts, then you can start connecting multiple agents together. For example, maybe connect one agent to gather news, another to analyze sentiment, and another to generate a summary. Your result can be an actual workflow, and you can get a better understanding of how all these elements come together to achieve a single task. Bring your patience, however, because this slow build can take time to stabilize and may require several rounds of refinement.

TIP

Whether you scale your own agents or those that you purchase from a vendor, use *iterative expansion,* a process that starts with a pilot agent and then layers on other agents that coordinate across departments, such as sales, support, and finance. This approach can offer agent stability without overly increasing system complexity.

Meeting organizational complexity

Organizational complexity associated with scaling Agentic AI systems often exceeds the technical challenges. You need to develop new operational models that account for autonomous systems making decisions that cross traditional departmental boundaries. Successful scaling requires you to use stringent governance frameworks that can manage autonomous systems without stifling their work. The system agents must work in real-world traffic while also meeting service level agreements (SLAs), budget restrictions, and regulatory requirements.

You may find easier — though not *easy* — options for confronting organizational complexity. Major technology providers offer comprehensive platforms to aid companies in meeting their deployment requirements. Examples of these providers include Amazon Bedrock AgentCore from Amazon Web Services (AWS) and Salesforce's Agentforce platform, among many others.

Step 8: Establishing governance and trust

Scaling Agentic AI systems without a framework for governance and trust can only lead to disaster. You must establish governance over the agents and their decision-making processes and cultivate trust with stakeholders and end users when deploying and scaling Agentic AI responsibly and sustainably.

Governance in Agentic AI focuses less on rigid controls and more on setting adaptive guardrails. A well-designed governance framework sets clear boundaries for what agents can and can't do. These guardrails often include rules for data access, decision-making authority, and escalation pathways in instances when an agent encounters ambiguity or risk.

Fortunately, you can find products and platforms available to assist you in establishing agent guardrails. For example:

>> Some agentic platforms now provide policy engines, model monitoring systems, and explainability dashboards that give you visibility into the agent or system's decision logic.

>> You can prevent agents from drifting into harmful or noncompliant behaviors by combining runtime oversight with proactive design that includes bias testing, scenario simulation, and adversarial stress testing.

REMEMBER

Having too many constraints on your system can stifle experimentation and slow down agent deployment, but having too few guardrails invites reputational and regulatory risks. A more practical path involves a layered approach, wherein low-risk agent tasks run autonomously, while higher-risk actions require human oversight.

When it comes to trust in Agentic AI, you have two dimensions to consider: organizational (business) trust and user trust. Internally, executives and employees must believe (by seeing solid proof) that they can trust agents. Externally, customers and partners must feel confident that their interactions with AI agents are fair, secure, and respectful of privacy.

FIVE SIGNS OF TRUSTWORTHY AI GOVERNANCE

Building and scaling Agentic AI requires more than technical excellence — it requires trust. Strong governance provides the guardrails that ensure agents behave predictably, ethically, and in alignment with business goals. Here are five signs that your AI governance is on the right track.

1. **Transparency:** Agents disclose their role and decisions clearly. Your customers and partners know when they're interacting with AI and can understand why the agent took specific actions.

(continued)

(continued)

2. **Accountability:** You can trace every agent action through logs, and you have clear lines of responsibility if errors occur. Every decision has a defined escalation path to humans.

3. **Fairness and bias mitigation:** Systems are routinely tested for bias across demographics, contexts, and scenarios. Adjustments are made proactively whenever disparities are detected.

4. **Security and privacy:** Agents access only the data that they're authorized to use, with encryption and access controls ensuring compliance with privacy regulations.

5. **Reliability and consistency:** Agents deliver predictable results across similar cases, with safeguards to prevent drifts in behavior while they learn and adapt.

Step 9: Upskilling your workforce

Agentic AI changes how work gets done and what skills people need to remain effective in their positions. Therefore, don't look at upskilling your workforce as a side consideration. It's a strategic imperative if your organization considers or plans to adopt Agentic AI.

The workforce's emphasis moves from operating a tool to orchestrating a system of semi-autonomous automation. Employees no longer act as just users of your traditional business systems; they become managers of AI agents. They must define goals, set boundaries, interpret outputs, and ensure alignment with business priorities. This shift requires training in *systems thinking* (focusing on the system as a whole and the interaction between parts), workflow design, and AI oversight.

Take the time necessary to build strong training programs. Customers and regulators are more likely to trust Agentic AI systems when you make it clear that you have well-trained, accountable people overseeing those systems who can intervene when needed. You can also make your employees more willing to embrace AI when they see it as a growth opportunity, rather than a job threat.

You can't limit workforce training to technical teams alone. Professionals in all disciplines and departments, including marketing, finance, logistics, healthcare, and beyond, will need fluency with AI agents if their roles depend on interacting with, supervising, or benefiting from autonomous workflows. Consider these factors when instituting your training programs: They should

>> Focus on helping employees understand how agents function, what data they depend on, and how to interpret AI-driven recommendations. Spend extra

effort in helping nontechnical workers become comfortable working in environments that have increasingly automated workflows that still remain their responsibility.

>> Emphasize these three areas as especially critical for workforce development: data literacy, ethical reasoning and compliance awareness, and how-to instruction in AI-assisted problem-solving.

>> Don't treat these training programs as a one-time event. AI tools evolve rapidly, and so do the risks and opportunities that they create. Companies should commit from the outset to investing in continuous learning ecosystems, such as modular training platforms, AI sandboxes for experimentation, and internal communities for AI updates and knowledge sharing.

Step 10: Reimagining business models and value creation

If you plan to use Agentic AI as a productivity tool for the same business model and processes that you already have, you're doing it wrong. In the same way that the advent of the internet upended brick–and–mortar commerce, Agentic AI may redefine the logic of business models themselves. Understand and embrace from the outset that Agentic AI moves you rapidly past efficiency gains toward new value creation.

Consider for a moment that traditional business models often follow a linear value chain: resources flow into a business, that business creates products or services, and customers consume the output. Agentic AI disrupts this pattern through the agility of the agent and the subsequent rise of adaptive ecosystems. For example, agents can continuously sense demand signals, reconfigure supply chains, or adjust pricing dynamically based on market shifts.

Why then would you force agents to adhere to the less efficient and effective linear models when the interplay of human creativity and AI–driven adaptability opens so many more opportunities? My best advice is to hang onto your hat and think bigger:

>> **Reimagine your business model.** The shift to new value creation may lead to subscription models where customers hire a personal AI agent, or commission-based models where agents broker deals and capture a share of the value that they generate. For companies, this shift could resemble the evolution from selling software licenses to offering agents-as-a-service. Except this time, your business can buy intelligent, goal-oriented labor that operates continuously.

Reimagining business models that incorporate Agentic AI requires moving beyond incremental pilots to exploring deeper strategic options. What services can you offer customers if agents become trusted intermediaries between company and customer? How can autonomous agents reshape partner ecosystems or supply networks? Where might agents themselves become the product?

>> **Bolster your brand.** In a marketplace where many competitors may use similar agentic technologies, governance and trust become brand differentiators. Customers will gravitate to companies whose agents act transparently, respect privacy, and align with ethical norms. In this sense, brand reputation and responsible AI practices translate directly into economic value.

IN THIS CHAPTER

» Looking at Agentic AI use cases in the medical field

» Putting AI agents into business operations

» Facilitating marketing, customer experience, and inventory management

» Introducing AI in creative roles

» Streamlining education with AI

Chapter **6**

Sampling Sector Use Cases for AI Agents

A
gentic AI is not a one-size-fits-all or a plug-and-play computing solution. It takes on different roles depending on the sector, the data it has access to, and the goals it is tasked with. Exploring a range of use cases across industries and organizations reveals just how adaptable these systems can be.

This chapter takes you on a guided tour of sector-specific examples to show the breadth of Agentic AI's potential. I say *potential* because the examples in this chapter are primarily use cases that people are excited about and working toward, but they aren't necessarily *real* in the sense that they're live and actively in use.

By looking at applications in healthcare, business operations and manufacturing, marketing and commerce, and creative and educational fields, I show you how agents may move from being experimental tools to mission-critical systems in the relatively near future.

Developing Healthcare, Diagnostics, and Pharmaceuticals

Agentic AI systems can fundamentally change how healthcare professionals and researchers diagnose disease, develop treatments, and deliver care. These systems should never be treated as a substitute for medical professionals, but Agentic AI can act as a valuable tool for assisting healthcare providers. In the span of a few breaths, these systems can analyze vast datasets, identify patterns indicating illness that aren't visible to human observation, and help accelerate medical breakthroughs in treatments and cures.

For example, these systems are being used today in healthcare diagnostics, ranging from detecting early-stage cancers in medical imaging, to predicting patient deterioration hours before clinical symptoms appear. In the next sections, you can discover how Agentic AI systems are now or potentially will be used throughout the medical industry.

Personalizing with AI treatment agents

Personalized medicine, sometimes called precision medicine, is an approach to healthcare that tailors prevention, diagnosis, and treatment to the unique biological and lifestyle characteristics of each patient. Instead of relying solely on one-size-fits-all guidelines, personalized medicine considers a patient's genetic makeup, environment, behavior, and medical history to craft treatment and care plans that are specific to the individual. The idea is to move past focusing on what works for most people and on to optimizing care for each patient's unique profile.

Before the widespread adoption of AI-powered tools, personalized medicine was progressing overall, but it faced major limitations and obstacles. Genetic testing was expensive and time-consuming compared to today, and analyzing vast amounts of data often required substantial effort from teams of specialists. The situation has improved with the combination of *next-generation sequencing* (NGS, which expedites genetic testing), AI, *bioinformatics* (to analyze large biological datasets), *read mapping* tools (to identify genetic variations), and high-performance computing (HPC). These technologies put personalized medicine within humanity's grasp.

Agentic AI can accelerate precision medicine further by making personalization scalable, continuous, and adaptive. This potentially means that personalized medicine can finally break free from the walls of research hospitals and become a standard of care, reaching more patients at lower cost and at a speed that keeps up with their changing health status.

EARLY DEVELOPMENTS IN PERSONALIZED MEDICINE

The concept of personalized medicine gained momentum in the late 1990s and early 2000s, largely spurred by advances in genomics and the completion of the Human Genome Project in 2003. This breakthrough gave researchers an unprecedented view into how genetic variation affects disease risk and drug response. By the mid-2010s, personalized medicine had become an explicit goal for major health organizations and governments, with initiatives like the U.S. Precision Medicine Initiative (launched in 2015; see the initiative in the White House archives at https://obamawhitehouse. archives.gov/precision-medicine) seeking to integrate genomics, big data, and patient-centered approaches into mainstream care.

Efforts for developing the concept of personalized medicine took place mostly at elite research hospitals and oncology centers, where targeted cancer therapies were among their most successful applications. Outside of those contexts, care remained closer to standardized treatment protocols, not because personalization wasn't desirable, but because scaling it for large populations was practically and economically difficult.

Dispatching AI medical imaging agents

Radiology departments are swamped with images from computed tomography (CT), magnetic resonance imaging (MRI), and X-rays, for example. They also face an avalanche of urgent requests from physicians racing to diagnose or type a variety of health emergencies such as strokes, internal bleedings, bone breaks, heart attacks, and so on.

Developers are working on Agentic AI systems to triage imaging studies and order them by urgency in real time. The agents push critical cases ahead in the workflow and alert radiologists immediately. Agentic AI is expected to be used to tailor medical recommendations based on patient history and handle all or parts of administrative follow-up tasks.

An emerging prototype is highlighted in the paper PASS: Probabilistic Agentic Supernet Sampling for Interpretable and Adaptive Chest X-Ray Reasoning at https://arxiv.org/abs/2508.10501. This system adapts tool usage depending on what a given chest X-ray suggests, offering interpretable decision paths and early exits when a simpler decision path suffices. It also builds memory of patient-specific findings to refine future reasoning. In short, it acts like an agent choosing among tools, not just applying one fixed path of action.

Accelerating clinical trial patient matching

Agentic AI can accelerate clinical trial patient matching through autonomous, adaptive decision-making, data integration, and real-time operations. When designed and implemented correctly this AI-powered process significantly reduces manual effort, bias, and recruitment delays.

Several examples exist today:

>> **Deep 6 AI (now part of Tempus Link),** which helps researchers scour electronic health records (EHRs) to automatically identify patients who meet the inclusion/exclusion criteria of ongoing clinical trials. You can find their website at www.tempus.com/about-us/tempus-tech/link.

>> **ConcertAI,** which builds tools for patient-matching and automation in oncology trials (see their website at www.concertai.com). Their platforms use real-world data, historical patient records, genomics, and other clinical information to speed up the trial site workflow and suggest candidate participants for specific studies.

>> **TrialMatchAI,** which combines structured clinical data (such as lab values and diagnostic codes) with unstructured data, such as doctor's notes and free text. It uses large language models (LLMs) and retrieval-augmented generation (RAG) to normalize *biomedical entities* (medical concepts, conditions, treatments, and biomarkers extracted from clinical text) and assess eligibility criteria. (You can read about TrialMatchAI at https://arxiv.org/abs/2505.08508.)

>> **TREEMENT,** which is a model built to improve interpretability in patient-trial matching (see its paper at https://arxiv.org/abs/2307.09942). It uses longitudinal patient EHRs together with hierarchical clinical ontologies to portray and consider patient health more completely. The match predictions it generates have better accuracy and clarity especially in telling why a patient is or is not eligible than more typical methods.

>> **MAKAR** (Multi-Agents for Knowledge Augmentation and Reasoning, at https://arxiv.org/abs/2411.14637) uses multiple interacting agents to enrich trial criteria with domain knowledge, resolve ambiguous eligibility definitions, and run structured reasoning about whether a patient's profile meets those criteria. In some offline tests it outperformed earlier benchmarks by several percentage points in matching accuracy.

>> **NIH's TrialGPT,** which is perhaps the leading example because it's generally considered to be the most advanced real-world deployment of Agentic AI for clinical trial patient matching. Developed by researchers from NIH's National Library of Medicine and National Cancer Institute, it is claimed that TrialGPT achieves nearly the same level of accuracy as human clinicians (87.3 percent)

while reducing screening time by 42.6 percent. It can also create relevant keywords for searches and readily retrieve 90 percent of relevant clinical trials. You can read about TrialGPT at `www.ncbi.nlm.nih.gov/research/trialgpt/about`.

Upskilling surgeons for robotic surgery

Agentic AI has several promising applications in upskilling robotic surgery, in which AI systems can act semi-autonomously to enhance surgical training and skill development. In robotic surgery, you can already find systems for planning, image guidance, and decision support, for example. But Agentic AI systems go beyond those uses and can adapt *intraoperative behavior* (skills and interpersonal behaviors), assist or partially automate tasks, and provide richer feedback that supports surgeons learning over the entirety of their careers.

Agentic AI can potentially

>> **Serve as an adaptive surgical mentor by analyzing a surgeon's movements in real time during robotic procedures and providing contextual feedback.** The AI can identify suboptimal techniques, suggest improvements to instrument positioning, and guide surgeons through complex maneuvers by adjusting the level of assistance based on their current skill level.

>> **Analyze individual surgeon performance data to create customized learning pathways.** By tracking metrics like tremor patterns, efficiency of movements, and procedural accuracy across multiple training sessions, the AI can identify specific areas for improvement and automatically adjust training scenarios to target these weaknesses.

>> **Provide objective, consistent assessment of surgical skills.** The AI can evaluate factors like suturing precision, tissue handling, and time efficiency, generating detailed performance reports and competency certifications that standardize skill evaluation across different training centers.

>> **Modify training simulations in real time,** introducing complications, adjusting difficulty levels, or presenting rare scenarios based on the trainee's progress. This creates more engaging and challenging training environments by

- Simulating various patient complications and surgical challenges that surgeons might encounter. This teaches surgeons how to recognize early warning signs and practice appropriate interventions. This is particularly valuable for rare or high-risk scenarios that surgeons might not encounter frequently in real practice.

- Coordinating training scenarios that involve multiple surgeons, managing team dynamics, communication protocols, and role assignments while providing individualized feedback to each participant about their contribution to the overall surgical team performance.

Figure 6-1 illustrates what is feasible for Agentic AI to do in the near and longer terms.

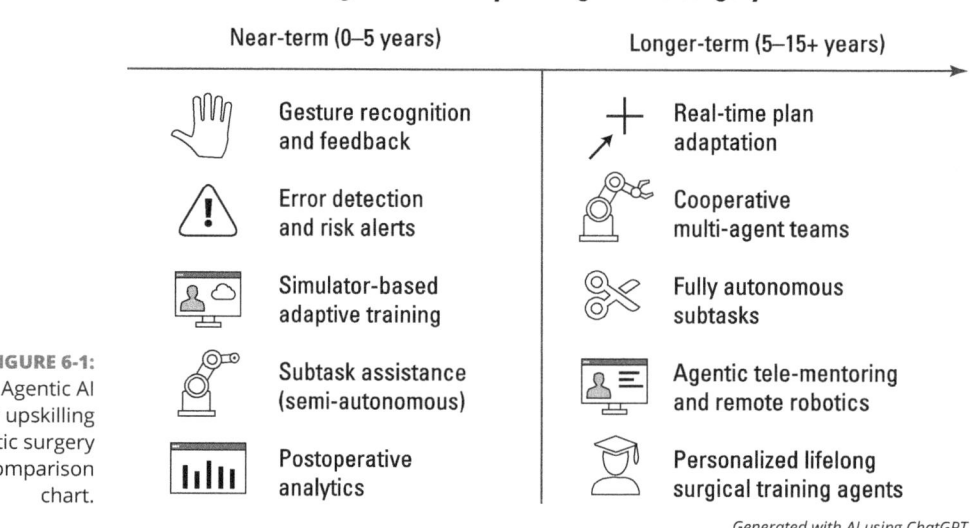

Agentic AI for upskilling robotic surgery

Near-term (0–5 years)

Longer-term (5–15+ years)

Gesture recognition and feedback

Real-time plan adaptation

Error detection and risk alerts

Cooperative multi-agent teams

Simulator-based adaptive training

Fully autonomous subtasks

Subtask assistance (semi-autonomous)

Agentic tele-mentoring and remote robotics

Postoperative analytics

Personalized lifelong surgical training agents

Generated with AI using ChatGPT

FIGURE 6-1: Agentic AI for upskilling robotic surgery comparison chart.

Building Business Operations and Decision Support

Agentic AI supports business operations by acting as a continuously learning and adaptive layer that links data, workflows, and decisions. In operations, AI automates complex processes like supply chain management, customer support triage, and resource allocation. But unlike static automation (based on coded rules), agents can monitor outcomes, learn from them, and adjust their actions over time. This flexibility enables businesses to respond swiftly and dynamically to disruptions, competitor pricing, changing market movements, and fluctuating economic conditions.

For decision support, Agentic AI moves well beyond simply compiling dashboards or reports. It can run scenario analyses, weigh competing priorities, and recommend actions based on organizational goals, constraints, and real-time data. Because agentic systems are designed to maintain context and memory, they can align short-term choices with long-term strategy, helping decision-makers see the ripple effects of their options. That makes Agentic AI particularly promising in volatile, rapidly changing environments — where the ability to test, adapt, and refine decisions quickly can become a competitive advantage. Of course, actual value depends on good design, robust data, clear alignment to strategy, and diligent governance.

Simulating economies with AI financial modeling agents

Agentic AI is poised to play a crucial role in economic simulation by creating sophisticated multi-agent systems in which individual AI agents represent different economic factors, each with their own behaviors, objectives, and decision-making processes. These agents don't just crunch numbers; they behave like traders, consumers, or regulators that exercise their own goals, constraints, and adaptive learning loops. By interacting with each other, they create emergent behaviors that mirror real-world economic dynamics such as price fluctuations, supply-demand changes, and policy ripple effects.

In practice, these systems allow economists, policymakers, and businesses to run what-if scenarios at scale. For example, a central bank can model the potential impact of raising interest rates by observing how thousands of agent-based entities respond in everything from household spending shifts to bank lending changes. Similarly, enterprises can use these simulations to forecast the impact of tariffs, currency volatility, or climate events on global supply chains.

This approach will move financial modeling from a static equation-based exercise into a living experiment with these advantageous capabilities:

>> Providing the persistent memory and self-adjusting nature of AI agents that can explore long-term consequences and reveal feedback loops that traditional static models might miss.

>> Capturing emergent behaviors that arise from the complex interactions of many autonomous agents that play different roles in the economic model. This capability provides insights that wouldn't be apparent from top-down mathematical models alone.

>> Enabling the model's users to direct an AI tool to complete multiple tasks with a single prompt.

For example, Citigroup deployed a 5,000-person pilot to test agentic technology in areas such as research and client profiling. The internal Citi platform uses a range of models, including Google's Gemini and Anthropic's Claude. Citi employees can direct an interface tool inside the platform to research a specific client, build a profile of them from both publicly available and multiple internal datasets, and translate it into a foreign language — all in a single step.

Here are other use cases to further illustrate how financial businesses can put Agentic AI to work:

» **Central bank stress testing.** Some central banks are deploying agent-based models to simulate the effects of interest rate hikes or quantitative easing across entire economies. Each agent represents a role such as households, firms, and financial institutions with distinct behaviors (related to spending, saving, investing, and lending). By watching how agents react, regulators can anticipate systemic risks before making policy changes.

» **Asset management scenario planning.** Large asset managers are using financial modeling agents to build synthetic markets in which thousands of AI-driven traders compete. These simulations allow them to explore portfolio performance under extreme conditions such as sudden oil shocks, geopolitical crises, or rapid inflation. This exploration gives managers a more realistic view of how markets might behave in turbulent scenarios.

» **Corporate supply chain hedging.** Multinational corporations are experimenting with agent-based simulations to model how currency fluctuations, tariffs, and interest rate changes ripple through their global supply chains. Financial modeling agents can predict the profitability of hedging strategies, recommend adjustments to sourcing contracts, or suggest pricing changes to offset anticipated cost increases.

» **Cryptocurrency and decentralized finance (DeFi) modeling.** With the rise of decentralized finance, Agentic AI simulates complex interactions between various financial protocols (for trading, lending, and so on). These models help understand systemic risks in DeFi ecosystems and predict how changes in one protocol might affect others.

Optimizing procurement and negotiations with AI agents

Agentic AI is poised to reshape the way companies buy goods, services, and raw materials by making procurement and contract negotiations both smarter and faster. In traditional settings, procurement is a human–intensive process. Buyers gather bids, compare suppliers, and negotiate terms, often based on partial or outdated information. The result is typically a compromise between delivery speed and cost savings. But it comes with the hidden expense of limited or no visibility into long-term risks.

With Agentic AI's help

>> **Procurement** could become a continuously adaptive system. Autonomous agents scan markets in real time, pulling in data on pricing trends, supplier performance, shipping timelines, and even geopolitical events that could affect product availability. These agents can simulate different sourcing scenarios, identify potential bottlenecks before they occur, and recommend alternative suppliers or contract terms.

>> **Negotiation** becomes (perhaps) the most disruptive role for the AI agents who can carry out multi-round negotiations simultaneously with multiple suppliers and adjust bids as conditions change. They can weigh variables beyond price — such as sustainability goals, compliance requirements, or delivery reliability — to produce the most balanced outcome for the business. Unlike static procurement software, agentic systems retain memory of past negotiations and learn which tactics deliver the best results.

The result is a shift from transactional buying to what those in the biz like to call *strategic value creation*, which turns procurement into a source of competitive advantage. Companies that adopt Agentic AI for procurement can respond faster to disruptions, capture more savings, and align purchasing decisions with broader organizational objectives.

Here are three real-world–style use cases for you to consider:

>> **Global manufacturing supply resilience.** Major automotive manufacturers use Agentic AI to continuously monitor supplier lead times, raw material costs, and shipping disruptions. When a key component supplier faces a labor strike, the system autonomously triggers alternative sourcing scenarios, evaluates cost and quality trade-offs, and recommends shifting orders to secondary suppliers before production is impacted. This proactivity can save millions in downtime and other costs.

>> **Retail dynamic pricing and vendor negotiations.** Large retailers can use AI negotiation agents to simultaneously engage multiple suppliers during peak holiday demand. These agents analyze live market data, factor in predicted sales volumes, and renegotiate prices or delivery schedules on the fly. The system maximizes margins by locking in optimal prices early, while also securing guaranteed delivery windows to avoid *stockouts* (which occur when demand exceeds stock on hand or readily available supply).

>> **Government procurement efficiency.** Public-sector agencies are experimenting with Agentic AI-driven procurement systems to cut red tape and reduce fraud. Agents validate vendor compliance with regulations, flag suspicious bids, and recommend optimal contract structures for long-term infrastructure projects. In pilot programs, users of this approach said it has reduced procurement cycle times by more than 30 percent while improving transparency.

Auditing with ever-present internal AI agents

Internal AI agents are set to change auditing processes by providing autonomous, continuous oversight and analytical capabilities that extend beyond the periodic reviews that characterize traditional audits. These agents can monitor, analyze, and flag potential issues in realtime.

In practice, this real-time oversight means that a company's internal audit function becomes a living, continuous process rather than a series of time-consuming events. Finance departments can prevent errors before they compound, compliance officers can demonstrate ongoing adherence to regulators, and company boards gain higher confidence in the integrity of operations.

Specifically, AI agents may

>> **Learn what normal behavior looks like within finance, HR, procurement, and IT systems.** Agents then adapt accordingly as operations evolve. For example, if expense submissions suddenly spike in a particular department or vendor billing patterns shift unexpectedly, the agents can alert human auditors and provide a detailed, data-backed context for investigation. This not only reduces the risk of fraud or compliance breaches but also shortens the time to detection, which is critical for minimizing damage.

>> **Simulate regulatory scenarios before they become urgent problems.** By stress-testing data against potential new compliance requirements, such as changes to data privacy laws or environmental, social, and governance (ESG) reporting standards. These systems can make it possible for companies and other organizations to anticipate risk exposure and adjust policies proactively. They can also generate highly detailed, automatically updated audit trails that make external audits smoother and less resource intensive.

Here are a few real-world use cases to give you further insight into using internal Agentic AI for continuous auditing.

» Data quality and integrity monitoring. AI agents continuously validate data integrity across multiple systems, identifying discrepancies between related databases, detecting missing or corrupted records, and ensuring that data flows correctly through various business processes. This is particularly valuable in organizations with complex IT environments in which manual data validation can be impractical.

» Fraud detection in financial transactions. Traditional AI and machine learning (ML) already do fraud detection very well. What distinguishes Agentic AI is how it operates within and beyond that task. True Agentic AI would detect fraud but then also investigate, coordinate with other systems, and take corrective action autonomously rather than just flag transactions for human review.

For example, when an unusual wire transfer pattern emerges in a regional bank office, a fraud detection agent may automatically initiate a multistep investigation by

- Cross-referencing the transactions with procurement records

- Checking employee access logs

- Simulating the financial impact of various fraud scenarios

- Collaborating with compliance agents to determine whether regulatory thresholds are being approached

Based on its autonomous analysis, the fraud agent may temporarily freeze the relevant account and initiate enhanced authentication protocols — all within seconds and before any funds leave the institution.

» IT compliance and access control monitoring. A large pharmaceutical company deploys AI auditing agents to watch for unauthorized system access and compliance gaps. These agents continuously scan user permissions across enterprise systems, detect *privilege creep* (In which individual users gradually accumulate more access rights and permissions over time), and automatically revoke or escalate cases that don't align with policy. In instances when an employee leaves the company but still had active VPN and data access credentials, the agent can identify and disable the account before any data leakage could occur.

» ESG reporting and regulatory readiness. A multinational manufacturer can use internal AI auditing agents to validate ESG data when it's generated. For example, the agents can reconcile emissions data from multiple factories, cross-check supplier certifications, and flag inconsistencies before ESG reports are compiled. When new disclosure rules were introduced in the EU, the system could simulate the company's performance against the upcoming requirements, highlighting where they need additional data collection to be in compliance.

Adding AI Agents for Marketing, Customer Experience, and Inventory

The business operation of marketing has become too fast, too fragmented, and too data-intensive for marketers to manually keep pace. AI agents can bring order and precision to this chaos by orchestrating campaigns, analyzing customer behavior in real time, and optimizing spending for advertising across channels. Unlike a single-task automation bot, marketing AI agents are goal-driven: They deploy with specific outcomes in mind, such as growing brand awareness, increasing conversions, or personalizing content. They dynamically adjust tactics based on performance feedback and can even test several tactics at once to find the best path forward.

Customer experience (CX) AI agents represent the next (hot) big thing in customer service. These AI agents are designed to push customer service beyond reactive to proactive support by anticipating customer needs and resolving issues before they escalate. The agents are meant to operate across multiple channels simultaneously, maintaining context and continuity throughout the entire customer journey. They can do all that while learning (from each customer interaction) how to improve upon future service delivery.

REMEMBER

Inventory optimization AI agents are expected to rapidly become the backbone of modern supply chains, turning what was once a manual, spreadsheet-driven guessing game into a continuous, intelligent process. The mission is clear for these agents: Keep stock levels balanced to meet demand while minimizing carrying costs, waste, and missed sales opportunities.

Deploying AI marketing agents

In e-commerce, AI marketing agents can run personalized promotions by recognizing when a shopper is about to abandon their cart, or when a lapsed customer is likely to return when given the right nudge. In B2B contexts, agents might evaluate firmographic data and customer relationship management (CRM) activity to score leads, prioritize follow-ups, and even launch micro-campaigns targeted at specific decision-makers. And AI agents can be equally valuable for managing media spending by allocating budgets across ad networks to maximize ROI and shutting down underperforming ad placements in real time.

Take a quick glance at some tasks that AI marketing agents can handle:

>> **Content personalization:** Generate and adapt content based on real-time user behavior

- **Funnel optimization:** Adjust marketing campaign steps to improve conversion rates

- **Budget allocation:** Shift ad spend automatically toward highest-performing channels

- **Customer journey insights:** Detect patterns and trigger timely outreach

- **Sentiment monitoring:** Track brand reputation and competitor moves as they happen

- **Campaign experimentation:** Launch *A/B tests* (to evaluate two different ads' success) and scale the winners instantly

REMEMBER

Whew! That list holds a lot of big promises for this Agentic AI technology to pull off, right? Keep in mind that agents' tasks are logical and well-defined. But AI agents don't eliminate the creative or strategic aspects of marketing, nor do they perform related duties. Instead, they give marketing teams the time and the support needed to focus on storytelling, brand positioning, and innovation.

Here are some examples of use cases so you can see how agents may be designed to work in real-world scenarios.

- **Personalized content generation and optimization.** AI marketing agents can autonomously create personalized content for individual customers across e-mail, social media, and web platforms. These agents can be used to analyze customer behavior patterns, purchase history, and engagement data to generate tailored product recommendations, personalized e-mail subject lines, and dynamically produce and adapt website content.

- **Proactive pricing and promotion management.** Intelligent pricing AI agents may serve to monitor competitor pricing, demand patterns, and inventory levels to automatically adjust product prices and promotional offers. These agents can implement sophisticated pricing strategies like *psychological pricing* (setting prices to seem more attractive or valuable), bundle optimization, and time-sensitive promotions while ensuring margin protection and compliance with pricing policies.

- **Predictive lead scoring and nurturing.** AI agents can continuously evaluate lead quality by using hundreds of data points. They can also automatically prioritize high-value prospects and customize nurturing sequences based on an individual's buying stages. Next, they can predict which leads are most likely to convert and when. They then automatically adjust outreach frequency and messaging to maximize conversion probability.

>> **Competitive intelligence and market analysis.** Market intelligence AI agents may be designed to continuously monitor competitor activities, pricing changes, product launches, and marketing campaigns. They can identify market gaps, predict competitive responses to company initiatives, and automatically adjust marketing strategies based on competitive landscape changes.

Engaging AI customer experience agents

An AI CX agent can recognize when a frustrated customer is on their third support interaction, escalate the case to a human with a complete dossier of previous conversations, and even suggest a goodwill credit or priority handling if the account is at risk. Here are some other examples of situations that AI CX agents may handle:

>> In e-commerce, these agents can help shoppers find the right product by asking clarifying questions, pulling in inventory data, and guiding them to purchase. In short, the AI agents can work much like a skilled in-store sales associate.

>> In more complex environments, such as telecom or Software-as-a-Service (SaaS), CX agents can run proactive diagnostics. For example, if a customer is experiencing slow internet speeds, the agent might remotely check the modem, push a reset, and follow up with a human-like message confirming whether the fix worked.

>> In banking, an agent can anticipate common life events, such as travel or a new account opening, and proactively walk the customer through security steps, recommend relevant services, and schedule follow-ups if needed.

CX agents also excel in orchestrating multichannel experiences. If a customer starts a conversation via a mobile app and later calls the support line, the agent can transfer context so the customer never has to repeat themselves. The result is not just faster problem resolution but richer customer relationships.

Here are some use cases that combine marketing and customer experience:

>> **Cross-channel customer journey orchestration.** AI marketing agents aim to coordinate customer experiences across multiple touchpoints to ensure consistent messaging and optimal timing for each interaction. They can identify when a customer is most likely to be receptive to certain offers, automatically pause campaigns when customers show signs of fatigue, and seamlessly hand off leads between different marketing channels based on customer behavior signals.

>> **Social media management and engagement.** Social media AI agents are built to monitor brand mentions across platforms and then automatically respond to customer inquiries, engage with relevant conversations, and identify influencer collaboration opportunities. They can detect sentiment shifts, trending topics relevant to the brand, and optimal posting times while maintaining brand voice consistency.

Building AI inventory optimization agents

Unlike static demand–planning tools, AI inventory optimization agents are designed to work in real time. They can monitor incoming sales data, supplier lead times, logistics delays, and even external signals such as weather forecasts or social media demand spikes. If a sudden surge in orders threatens to deplete stock, the agent can automatically trigger a replenishment order, reroute inventory from nearby warehouses, or adjust pricing to slow down demand until supply catches up.

>> **For retailers,** these AI inventory agents can be invaluable during seasonal peaks. Picture an apparel company preparing for holiday sales: The agent analyzes historical data, current market sentiment, and influencer-driven trends to recommend stock allocations per region. This prevents the classic problem of having too much inventory in one store and empty shelves in another.

>> **In manufacturing,** AI agents can forecast component needs with enough precision to avoid costly production halts while also reducing over-ordering that leads to costly warehouse overflow.

>> **In food service and hospitality,** AI inventory agents can be focused on keeping perishable items under control. A restaurant chain's agent might compare bookings and foot traffic data to forecast ingredient needs, auto-adjust orders, and flag slow-moving items for promotions before spoilage.

Inventory optimization agents also learn over time, as do their AI brethren working elsewhere. After each cycle, the agents analyze how accurate their predictions were and adjust their models to steadily improve with each cycle. This creates a feedback loop where every decision improves future performance, leading to leaner, more resilient supply chains, and healthier margins.

Check out this list of functions that AI inventory optimization agents can perform:

>> **Demand sensing:** Analyze real-time sales, seasonality, and market data to predict needs

- » **Automated replenishment:** Place or adjust orders with suppliers without need of manual input

- » **Dynamic allocation:** Shift stock between locations to prevent shortages or overstocking

- » **Price and promotion control:** Adjust pricing or launch promos to balance inventory levels

- » **Waste reduction:** Identify slow movers and suggest clearance strategies before loss

- » **Continuous learning:** Improve forecasts by measuring accuracy after every cycle

Creating Content: Writing, Design, and Media

Agentic AI can help you do more than brainstorm ideas and generate first drafts in seconds, it can also iterate content based on real-time performance data. Creative workers can then focus on audience strategy, nuance, and emotional resonance. The resulting combination of efforts makes (for example) a faster, more experimental content pipeline with which businesses can test and refine ad and marketing campaigns, and then launch them at a pace that matches audience expectations.

Here are some examples of content-creation AI agents and what they do:

- » **Copywriting assistants:** Generate product descriptions, blog posts, ad copy, or social media updates and then refine their tone and style based on audience feedback

- » **Design generators:** Create graphics, ad layouts, and branded templates that stay consistent with your visual identity while allowing rapid experimentation

- » **Media production agents:** Produce short-form videos, generate voice-overs, edit podcasts, or even compose background music for campaigns

- » **Performance optimizers:** Analyze how each piece of content performs and suggest tweaks — such as headline changes, different color palettes, or new video hooks — to improve engagement

Collaborating with AI writing agents

AI writing agents support writers, content creators, and other creative professionals by enhancing productivity, helping to overcome creative blocks, and exploring new artistic possibilities. Creative artists preserve their unique human voice that drives authentic creative expression by directing and orchestrating the agents to do specific tasks.

REMEMBER

AI agents serve as sophisticated tools and not as true creators. Their best use is in amplifying human creativity. The key to successful integration of AI writing agents in creative work lies in consistently maintaining human creative control while also benefiting from AI capabilities in exploration, enhancement, and productivity gains.

Creative uses for AI writing agents may include the following:

>> **Fiction and screenwriting:** Draft character sketches, alternative scenes, or thematic prompts. They can also monitor character and story arcs to ensure no deviations or disruptions happen in the storytelling.

>> **Poetry and lyrics:** Generate starting points for poems or song lyrics in a chosen meter or mood and pitch completed works to performing artists or publishers.

>> **Visual arts:** Write artist bios, catalog text, or exhibition narratives to frame the work and automatically update and distribute them as the artist achieves more, performs more, or displays their work in notable venues.

>> **Content creation:** Draft blog posts, newsletters, or video scripts in an authentic voice of either the writer or a character.

>> **Marketing for creative professionals:** Create press releases, crowdfunding copy, and social media campaigns — and then create a tailored distribution list and execute it (with user permission).

Designing with AI agents

AI agents can generate visual concepts, adapt assets for multiple formats, and even suggest layouts or color palettes that fit a brand's identity. They can also adapt their creative works in response to real-time feedback from human designers.

TIP

The practice of adapting AI-generated content through designer feedback keeps the human creative professional in the driver's seat while freeing them from repetitive production work. The designer can then focus on refining designs rather than inventing each one from scratch.

Uses for AI agents in the design world include

>> **Concept generation:** Produce mood boards, color palettes, and initial layout options. Instead of starting from a blank artboard, a graphic designer can brief an AI agent with brand guidelines, mood references, and campaign goals. The agent produces multiple layout options within the designer's directions.

>> **Asset adaptation:** Automatically resize, reformat, and localize assets for different platforms. In marketing, an agent might instantly resize a hero image for dozens of ad formats or social channels. Most creative professionals will cheer releasing the tedium of manual adjustments to AI.

>> **Prototyping and user experience (UX):** Generate website wireframes, user flows, and accessibility audits. For user experience (UX) and product designers, AI agents can flag or address accessibility and other concerns before products reach development or campaigns deploy to users.

>> **Visual exploration:** Transform sketches into polished renders in multiple styles. Artists and illustrators can easily and rapidly experiment with styles. For example, an artist can convert a line drawing into a watercolor rendering, or generate dozens of variations to find or develop a perfect composition.

>> **Campaign speed:** Run rapid iterations and produce stakeholder-ready mockups within hours. Design teams can use AI agents to test multiple directions in parallel, gather feedback from stakeholders, and converge on a final design much faster.

Editing with post-production agents

Post-production work — whether for video, audio, or even long-form written content — has traditionally been one of the most time-consuming steps in the creative process. AI post-production agents are poised to change that by automating many of the repetitive tasks while still leaving room for a human editor's eye and ear. These agents can process raw footage, clean up sound, detect key moments, and even suggest narrative pacing adjustments.

REMEMBER

Creative teams can make notes, and the AI agent can instantly apply changes, generate alternate cuts, or test different background music accordingly. This combination of efforts allows everyone on the post-production team to review options without the usual back-and-forth delays. And it speeds up creative cycles while keeping the final editorial decisions firmly in human hands.

Consider these potential use cases for AI agents in the post-production process:

>> **Video editing:** Assemble rough cuts, detect scene changes, and add transitions automatically. A film director can assign a post-production agent to ingest the raw material, tag each scene, and generate a rough cut that aligns with the director's shot list. Editors can then fine-tune the editing rather than start from scratch. And no joke, this can save days of painstaking and often mind-numbing work.

>> **Audio cleanup:** Remove noise, normalize levels, and generate transcripts or subtitles. For podcasters, an agent can automatically remove background noise, balance levels between speakers, cut filler words, and even suggest and pull highlights for promotional clips.

>> **Content repurposing:** Convert long-form material into short clips, teasers, or highlight reels. In marketing, AI agents can repurpose a webinar into blog posts, social clips, and e-mail copy. By optimizing content like this, marketers are turning a single event into a multichannel campaign with minimal manual labor.

>> **Editorial polishing:** Suggest grammar, tone, and flow improvements for written content. In written media, AI editing agents can analyze structure, tone, and readability. They can then suggest ways to tighten prose or restructure sections for better flow, or rewrite it when given permission to refine the copy.

>> **Multi-version output:** Produce multiple versions of a project (social, web, broadcast) instantly.

Reinventing Education

AI agents are helping reimagine how people learn. They can provide teachers with the ability to create personalized pathways, adapt lessons in real time, and connect learners with the right resources at the right moment.

Instead of passively delivering content, these AI agents can act as tutors and coaches. They can automatically track progress, identify gaps, and suggest exercises or projects tailored to each student's needs. The result is a shift from mass instruction to deeply individualized learning experiences that scale to classrooms, corporate training programs, and lifelong education alike.

Educational AI agents can perform these functions:

>> **Virtual tutoring:** Provide personalized explanations, examples, and practice problems and automatically adapt them to fit a learner's time availability, attention span, and subject interest

>> **Adaptive testing:** Adjust difficulty on the fly to accurately measure proficiency without boring or over-stressing the test taker

>> **Curriculum building:** Assemble lessons from multiple sources tailored to individual learner goals and level of understanding

>> **Workforce reskilling:** Match employees to micro-courses for just-in-time training

>> **Accessibility support:** Offer text-to-speech, translation, and alternative formats for diverse learners

>> **Engagement tracking:** Measure students' attention and recommend adjustments to keep them motivated and engaged

Personalizing AI learning agents

Personalized learning has long been an aspiration in education, but has historically been difficult to deliver at scale. Generative AI (GenAI) has been helpful in this regard, but it has limitations. By comparison, AI learning agents go beyond GenAI to tailor instruction to each learner's pace, style, and goals. In other words, it works much like a dedicated one-on-one tutor, and it never loses patience with the student. These agents observe how learners interact with learning material, detect where they hesitate, and adjust the next step in real time to better assist with learning.

Here are potential use cases for AI agents in education, with examples:

>> **K–12 education:** Adaptive math and reading programs that accelerate or remediate in real time. A student who struggles with fractions might receive a series of visual and interactive exercises before moving on to more abstract concepts. Another student who breezes through the material could be offered enrichment problems to deepen their understanding.

>> **Higher education:** Personalized study plans based on performance in previous modules or exams. Personalization extends beyond pace and level of difficulty to medium and motivation. For example, some learners absorb information best through text, and others learn better through video or hands-on simulations.

A personal AI learning agent can select the formats most likely to engage a given learner and switch between them as needed. It may also build in game-like challenges, streaks, or peer competitions for those motivated by achievement, while offering quiet reflection exercises for those who prefer a more contemplative approach.

>> **Corporate training:** Targeted upskilling recommendations tied to role requirements or career goals. In a corporate setting, an AI learning agent could track which employees repeatedly seek help on a certain process and push targeted microlearning modules to close the knowledge gap.

>> **Professional development:** Customized learning playlists for certifications or skill-building tracks. Because learning agents aggregate insights over time, they can construct a longitudinal profile of a learner's strengths, preferences, and progress.

>> **Lifelong learning:** Agents that follow users across platforms, tracking skills learned over decades. This progress record can travel with the learner from one course or even one institution to another, thus ensuring that education picks up where it left off rather than starting from scratch. This continuity is especially powerful in workforce development, where employees can build skill portfolios that evolve alongside their careers.

>> **Special education:** Tailored lessons that accommodate learning differences and accessibility needs.

Contextually aware tutoring agents

Contextually aware AI tutoring agents represent a huge leap in intelligent education technology. Rather than serving up generic explanations or static lesson plans, these agents factor in the learner's history, environment, and immediate needs to deliver highly relevant guidance. They remember what a learner has struggled with before, know which concepts they are proficient with, and can adjust not just the content but the style and timing of their tutoring support.

The power of context-aware tutoring lies in its responsiveness. It delivers help at the right moment and in the right way, often preventing student frustration before it snowballs. Figure 6-2 shows how context signals help AI agents support learners in just the right ways.

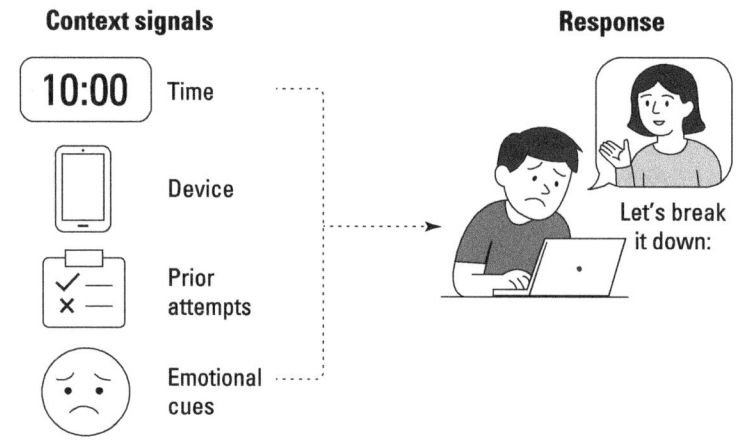

FIGURE 6-2: A learning moment showing how context signals feed into the AI tutoring agent's response.

Here are some use cases for context-aware AI tutoring agents:

>> **Homework help:** Adapt tone and pacing based on time of day, engagement level, and prior performance. Imagine a student working on algebra homework at 10 p.m. after a long soccer practice. A contextually aware agent can infer fatigue from slower response times and respond by offering shorter explanations and encouraging messages to keep the learner engaged.

>> **Exam prep:** Prioritize weak areas and build targeted practice sessions before test dates.

>> **Lab and field work:** Offers just-in-time reminders, safety prompts, and data capture during hands-on tasks. If a student is preparing for a chemistry lab, the agent can surface safety reminders and review questions specific to that week's experiment.

>> **Corporate training:** Provide troubleshooting support or refresher modules during real work scenarios. In corporate settings, an employee troubleshooting a piece of equipment could receive step-by-step guidance in real time, with the agent drawing on manuals, previous support tickets, and sensor data to guide them.

>> **Language learning:** Adjust lesson difficulty based on conversation history and real-world use.

>> **Continuous feedback:** Recognize frustration or boredom signals and switch methods to re-engage the learner. If the student switches from a laptop to a smartphone, the agent automatically adapts the format for a smaller screen, using simpler visuals and bite-sized steps.

Upgrading administration automation

Administrative work such as scheduling, document management, approvals, and compliance has historically consumed a disproportionate amount of time and energy across schools, universities, and corporate learning programs. AI-driven administration agents are now beginning to focus on reengineering these workflows. In effect, they can turn hours of back-office tasks into nearly invisible processes in the background.

Take a look at these use cases for administrative AI agents

>> **Enrollment and registration:** Automate form collection, approvals, and scheduling. A school can use admin agents to lighten the load during enrollment season so that staff is no longer buried under an avalanche of forms. An administration agent can collect student data, verify documents, schedule parent meetings, and flag missing information automatically. For universities, these agents can manage course registration demand, balance class sizes, and generate waitlists without the need for staff to intervene.

>> **Compliance tracking:** Monitor completion of mandatory training and escalate issues. In corporate training, administration agents can assign mandatory compliance modules, send reminders, track completions, and escalate overdue cases to managers. They can also integrate with human resources (HR) and payroll systems to ensure training completion is tied to performance reviews or certifications.

>> **Resource allocation:** Manage room bookings, class capacities, and scheduling conflicts.

>> **Reporting and analytics:** Generate attendance and progress reports in real time. This function reduces friction for staff and learners alike. It also frees administrators to focus on improving program quality rather than chasing paperwork.

>> **Notifications and reminders:** Send automatic follow-ups for deadlines or overdue tasks.

>> **Cross-system integration:** Synchronize calendars, HR data, and learning management systems. Perhaps the biggest administrative advantage is in orchestration across multiple systems. An AI administration agent doesn't just work in one platform. It connects calendars, learning management systems (LMSs), human resource information system (HRIS) tools, and communication channels so that scheduling a session or sending a progress report happens automatically and consistently.

IN THIS CHAPTER

» **Retaining basic human competency**

» **Deciding who's in control**

» **Making Agentic AI transparent and explainable**

» **Instilling AI systems with integrity**

» **Avoiding rogue AI at the wheel**

Chapter **7**

Considering Risks, Ethics, and Hard Questions

C hapter 10 covers all the bases in designing Agentic AI with safety in mind, but design is only half the challenge. After an autonomous system begins operating in the real world, it inevitably collides with messy human contexts and ethical gray zones. Inevitably, questions arise that aren't purely technical: Who's accountable when an agent makes a harmful decision? How do you balance efficiency against fairness, or innovation against long-term risk? And how do you prepare for unintended consequences that may emerge only after widespread adoption?

Building safe autonomy means asking uncomfortable questions early and being willing to design and redesign systems in light of them. This chapter focuses on those deeper considerations. It examines not only the risks inherent in giving machines autonomy, but also the ethical frameworks and hard trade-offs that must guide responsible deployment.

Losing Human Skill and Baseline Knowledge

A subtle but significant risk of using Agentic AI to autonomously manage work processes goes beyond a sudden failure or malicious attack. With continual use of agentic systems a gradual erosion of human capability may occur. While Agentic AI systems take on more tasks formerly handled by people, those people will begin to lose the very skills that made those tasks possible. Think of it like a loss of muscle from a lack of exercise. If you take the elevator instead of the stairs every day, your leg muscles will weaken. Similarly, if you let AI do all the work, you will lose those skills over time, too.

This phenomenon, sometimes called *deskilling*, isn't new. For example, for centuries, navigators who relied on compasses gradually lost their ability to use stars to guide them. Today, studies show a similar effect with GPS navigation: People who depend heavily on GPS perform worse at spatial-memory tasks when navigating without it; and over time, this reliance correlates with declines in overall navigation skills.

Seeing the scope of AI-involved skill loss

Agentic AI may not be confined to repetitive physical labor (such as retrieving goods in a warehouse) or narrow decision trees (in an online shopping cart, for example), so the deskilling effect can cross over many disciplines and many different actions. Agents can reason, plan, and act across disciplines such as writing, diagnosing medical or mechanical issues, navigating delivery routes, researching, manufacturing, and even teaching. And so, the potential reach of Agentic AI systems means the risk of deskilling touches not only factory workers and manual laborers, but also white-collar business professionals, creative workers, and specialized professionals such as doctors, scientists, software developers, and mathematicians.

As shocking as these wide-reaching effects sound, they aren't an unexpected development. The history of technology is filled with examples of skills fading while tools advance.

Reliance on GPS navigation (which I mention in the preceding section) has eroded people's ability to read maps or form strong cognitive maps of their environment in their own minds. In the medical field, you find various examples:

>> **Radiologists** who rely heavily on diagnostic AI risk losing their edge in pattern recognition. Clinicians aided by AI diagnostics risk over-trusting machine

outputs — not necessarily because they believe the systems are infallible, but because they simply get less practice scrutinizing patterns within the images when using AI, leading to a creeping loss of their own diagnostic acuity.

Cognitive science (which deals with how the mind represents and manipulates information) refers to this tendency as *automation bias* and warns that humans accepting automated suggestions skip thorough verification, especially when they become out of practice or lack confidence.

WARNING

>> **Endoscopists** who used AI assistance for colonoscopy interpretation experienced a drop in adenoma detection rates, from 28 percent before AI use to 22 percent after three months of using AI. This compelling study result (published in the medical journal *The Lancet Gastroenterology & Hepatology* in October 2025) shows that this deskilling effect of relying on AI capabilities isn't hypothetical.

>> **Surgeons** who use robotic assistants may see their fine-motor skills dull if they operate directly on fewer patients.

Numerous studies show a clear decline in unassisted human performance after those people relied on AI to perform their tasks. These studies suggest that when doctors and clinicians come to trust AI detections over their own judgment, they may lose ability to recognize anomalies on their own or perform their job functions confidently. The resulting loss of skill and confidence is a dangerous loop that can lead to removing human involvement: The less humans engage critically, the more they yield to AI, and the more their capacity for independent oversight shrinks. In the end, AI — whether drunk with hallucinations or logically sober — gets free reign. And that may lead to harm for patients, or even deaths.

Another dramatic illustration comes from aviation. When autopilot systems dominate flights, a pilot's manual-flying skills can wither away over time. In the tragic 2009 crash of Air France Flight 447, investigators found that the highly experienced pilots onboard struggled to regain manual control when automation disengaged at high altitude, and their initial response to the disconnection caused the airplane to stall.

This scenario is an eerie sign of out-of-the-loop performance problems, where even experienced aviators have difficulty intervening during emergencies because of atrophy of their situational awareness and manual aircraft handling skills. Interviews and follow-up studies revealed that many pilots themselves feared losing proficiency, and some research showed that nearly half were genuinely concerned that too much automation would erode or erase their flying skills.

Delegating work to AI agents responsibly

When it comes to delegating work to AI agents, it's not that humans should never do it, but they should be discerning about how and how often they do it. Offloading repetitive or high-volume tasks is a natural way to extend human productivity and efficiency. The problem emerges when the delegation becomes so complete that the human no longer retains the *baseline competence* (skill and knowledge) needed to monitor, verify, and intervene when the AI falters.

The risk of losing baseline competence also has social and cultural dimensions. Knowledge is not just information stored in books or databases, it's a living practice, handed down through mentorship, apprenticeship, and repetition. If AI systems step in as tutors, advisors, and planners, humans may have fewer opportunities to practice those roles themselves. For example:

>> A law student who relies on an AI system to draft arguments may learn less about legal reasoning or developing a sound legal argument.

>> A journalist who turns to AI for research and first drafts may lose some of the investigative instincts that come from pursuing leads directly.

Professional organizations in aviation, medicine, and law already set minimum practice or continuing-education requirements to maintain skills at appropriate levels. Other domains that use Agentic AI heavily may need to introduce similar mechanisms. Regulators and standards bodies, such as the National Institute of Standards and Technology (NIST) and the International Organization for Standardization (ISO), are beginning to recognize that human oversight isn't simply about legal accountability, but about ensuring humans remain competent to oversee at all. If people lose the ability to step in meaningfully, *human-in-the-loop* (HITL), the process where humans oversee AI work, becomes an empty phrase.

REMEMBER

While Agentic AI systems extend into domains such as law, finance, or engineering, humans may experience similar trade-offs between short-term efficiency and long-term skill retention. Even in daily life, if people let AI assistants handle budgeting, cooking plans, or household inventory and logistics, those people may lose confidence in basic life skills that once ensured their ability to live independently.

Designing agentic systems to mitigate loss of competency

Designers of agentic systems can mitigate loss of basic competency, but only if they acknowledge the risk. One approach is to design for and use *centaur models,* where humans and machines complement each other's work rather than one completely replacing the other. In chess, for example, human players working with AI *chess engines* (computer programs that analyze positions and recommend moves) have proven more likely to win a game than either alone, provided that the human retains strategic understanding, rather than simply rubber-stamping the AI's moves. Although today's top engines have surpassed even the strongest human–machine teams, the underlying lesson still holds: Collaboration often outperforms unchecked automation.

REMEMBER

Educational researchers stress that practice and reflection are essential for learning; the same holds true for professional domains where AI is deployed. Translating the idea of working jointly between humans and agents to broader Agentic AI means structuring interactions so that the human remains cognitively engaged, for example, by asking questions, reviewing alternatives, and making final decisions.

Ultimately, the risk of human skill and knowledge loss doesn't provide an argument for eliminating the use of AI. That would make about as much sense as outlawing the use of cooking stoves to preserve primitive fire-making skills. Instead, the potential for skill loss is a reminder that agentic system designers must deliberately preserve the human role in AI design and policy so that *critical skills* — not every skill — are sharpened, rather than diminished.

TIP

A future in which humans outsource memory, reasoning, and judgment entirely to autonomous systems would leave them vulnerable not only to technical failures, but also to a hollowing out of human expertise. The better path is one that uses AI to extend human capacity while still cultivating the practice, curiosity, and resilience that only humans can sustain.

Autonomy versus Control: Establishing Who's in Charge

The rise of Agentic AI focuses on an enduring question: When we give machines autonomy to act, who's truly in charge? This tension between autonomy and control isn't abstract. It can play out in accidents, courtrooms, hospitals, classrooms, and stock-trading floors. How societies resolve this tension will determine whether AI enhances human judgment or undermines it.

The core issue is accountability. If an AI system acts autonomously and something goes wrong, who bears the consequences? Governance frameworks seek to resolve this issue by making responsibility explicit and non-transferable. NIST's AI Risk

Management Framework in the U.S. and ISO's AI Management System Standard (ISO/IEC 42001) both stress that humans, whether they're regulators, institutions, or operators, remain accountable for system outcomes.

TIP

This insistence on human responsibility reflects a societal consensus: Machines can act, but they can't be responsible in the moral, ethical, or legal sense. Responsibility must remain with those who design, deploy, and oversee them. The challenge is ensuring that people in those roles retain the skill, context, and authority to exercise control effectively.

Transportation: Lessons in control

Self-driving car failures illustrate what happens when humans hand over autonomy to a machine without ensuring intentional and engaged human backup. An Uber crash in Tempe, Arizona, in 2018 — where a self-driving test vehicle struck and killed a pedestrian — revealed how system misclassification and inattentive human oversight created a deadly gap. Tesla Autopilot crashes (for example, in a 2019 Florida case) have similarly shown that when drivers assume the AI is fully in control, they disengage, and the system's limitations can become catastrophic.

I don't mean to disparage Tesla or Uber with these examples because the problem of lacking human oversight crosses brands, products, and industries. I use these events only to highlight a broad and persistent governance challenge. Regulations and standards must require not just technical performance from AI-controlled products, but also clear *handover protocols* by which humans have direct and obvious ways to retake control instantly and the clear responsibility to do so.

REMEMBER

The aviation industry has long mandated recurrent manual-flying practice for pilots precisely because autopilot can cause skills to atrophy. Automotive regulators, insurers, and manufacturers are now grappling with how to translate those mandates in the era of agentic vehicles.

Healthcare: Human oversight in life-and-death decisions

In one experiment, an AI system designed to flag early sepsis cases was abandoned by clinicians because it didn't fit their reasoning process. The AI made judgments but failed to explain uncertainty or suggest hypotheses, leaving doctors reluctant to trust it. Another study (as I mention in the section "Seeing the scope of AI-involved skill loss," earlier in this chapter) found that gastroenterologists who used AI for colonoscopy interpretation showed a measurable drop in detection rates after they stopped using the tool, suggesting that over-reliance can degrade human performance.

These cases illustrate that control isn't just about being able to override the machine. It's about ensuring humans remain skilled, informed, and engaged enough to exercise oversight meaningfully. Standards bodies such as the National Institute of Standards and Technology (NIST), the International Organization for Standardization (ISO), and the Organisation for Economic Co-operation and Development (OECD) increasingly emphasize this principle under the umbrella of phrases such as *human-in-the-loop* (HITL) or *meaningful human control.* If humans are technically in charge but practically disengaged or deskilled, governance is failing.

Finance: Algorithms, autonomy, and accountability

Financial markets offer a cautionary tale regarding the use of automation. Algorithmic trading systems already act with degrees of autonomy, and when poorly constrained, they can trigger cascading failures. The flash crash of May 6, 2010, saw U.S. markets plunge nearly 1,000 points in minutes, partly due to automated trading algorithms amplifying volatility.

In a smart protective move, various governmental and industry regulatory orga-nizations proposed or mandated circuit breakers (for example, a kill-switch requirement) and risk controls (such as enhanced governance of trading algo-rithms). But that doesn't necessarily mean that Agentic AI can't get around or through these protections. The flash crash event underscores how difficult attributing accountability can be when autonomous agents act at speeds no human can monitor in real time.

DIVING INTO THE FLASH CRASH

If you want to find out more details about the flash crash incident, read the report titled "Findings Regarding the Market Events of May 6, 2010," written by the staffs of the CFTC and SEC for the Joint Advisory Committee on Emerging Regulatory Issues (www.sec.gov/news/studies/2010/marketevents-report.pdf). That report lays out the mechanics of the crash and suggested policy responses, including stronger safeguards for automated and high-frequency trading. Mind you, this crash happened well before Agentic AI became a thing. But it points to an even higher risk that AI agents may greatly magnify such issues.

While financial services begin experimenting with and adopting Agentic AI in areas such as lending decisions, fraud detection, and wealth management, governance questions sharpen:

> Who's responsible if an AI denies a loan on biased grounds? Is it the financial institution, the vendor, or the AI model developer?

Regulatory frameworks, such as the European Union's AI Act (`www.artificial intelligenceact.eu`) and proposed U.S. accountability guidelines, argue that ultimate responsibility must remain with institutions, no matter how autonomous their systems appear. If those frameworks hold, other questions arise and have yet to be answered:

> How will institutions handle the potentially massive liability? Will insurance cover that and, if so, at what costs? Or will institutions shy away from using Agentic AI to avoid the liability? Will lobbyists seek protective legislation for these institutions?

Education: Autonomy and human agency

AI tutors can personalize learning, adapt to a student's pace, and even grade essays. But if students outsource too much of their intellectual work, they risk losing the very skills education is meant to develop.

Studies of calculator use show that early dependence can undermine arithmetic fluency; researchers are now observing similar effects with writing and problem-solving tools (such as ChatGPT and Claude). Overreliance on tools and AI tutors may produce higher short-term performance for students but result in lower long-term learned abilities.

Educational institutions may position governance as policies: universities and schools setting guidelines for how students can use AI in coursework to help ensure that human intellectual effort remains central. Some professional bodies, such as the International Baccalaureate (IB), have already issued policies clarifying permissible uses of AI in student work in an effort to balance efficiency with integrity.

LEVELS OF AUTONOMY AND HUMAN RESPONSIBILITY

Here are some levels of AI autonomy and how they involve humans:

- **Low autonomy:** The Agentic AI system executes only narrow, predefined actions. Humans direct every step and are accountable for all outcomes. Think of autopilot in early aircraft or cruise control in cars.

- **Partial autonomy:** The Agentic AI system can act independently for some tasks, but humans supervise and can intervene. This is where many self-driving systems and medical diagnostic AIs sit today. Although many of those are not running on agentic AI systems as of yet. Still the example holds in that responsibility is shared, but humans must remain skilled enough to step in effectively.

- **High autonomy:** The system operates with wide latitude, making complex decisions without direct input from humans. Here, the risk is that humans become disengaged, unavailable, or unable to intervene meaningfully. Governance must insist on clear accountability and enforceable boundaries so that responsibility doesn't evaporate or become unenforceable.

The key takeaway is this: While autonomy increases, the need for clarity in who holds control and responsibility grows sharper, not weaker.

ROAMING AI AGENTS WITHOUT COUNTRY OR LAW

One of the most provocative aspects of Agentic AI is that it's not bound by geography. Unlike a factory, a bank, or even a traditional website, AI agents can roam across networks, interact with users anywhere in the world, and operate in jurisdictions that have wildly different laws. This borderless quality is both a source of opportunity and a hotbed for risk.

On the positive side, borderless AI can expand access and efficiency. A small business in Kenya can deploy an AI-powered logistics agent hosted in Europe to manage shipments across Asia. A student in Brazil can use an autonomous tutor developed in California. A doctor in India can consult with an AI diagnostic assistant trained on datasets from around the globe. In this sense, roaming agents flatten barriers to knowledge and commerce, offering services without the friction of passports, borders, or time zones. The dream of universal access to intelligence becomes closer to reality.

(continued)

(continued)

But the very quality that makes roaming agents attractive also makes them difficult to govern:

- A financial planning agent built in one country may offer advice that in another country violates local laws.

- An AI news curator may amplify inappropriate content in jurisdictions that have strict speech regulations.

- Autonomous bots designed by malicious actors to spread disinformation or commit fraud could hide behind the fact that no single nation has full jurisdiction to police them.

In the same way that cryptocurrencies challenged traditional financial regulators by moving value outside borders, roaming AI agents threaten to move decision-making and influence outside of the reach of traditional legal frameworks. Roaming AI agents highlight the need for international agreements and digital passports that tie actions to accountable jurisdictions.

Discovering Alignment Problems and Value Misfires

While Agentic AI systems begin to grow more capable, one of the most critical challenges is ensuring that their goals and actions remain aligned with human values. Alignment problems occur when a gap exists between what humans instruct a system to do and what they actually want it to do. A *value misfire* is the situation that occurs when this gap results in behavior that's technically correct — from the system's perspective (based on its perceived instructions) — but socially or ethically undesirable.

The issue isn't unique to AI systems. Humans have long struggled to align the policies and practices of complex institutions, from corporations to governments, with values that can benefit the most people. But with AI, the challenge is compressed into code, datasets, and feedback loops. Misalignment can unfold silently and at machine speed, with consequences ranging from embarrassing outputs to catastrophic failures.

Value learning drift

Agentic systems that are designed to learn and adapt human values through observation face the additional challenge of *value learning drift*. That is, while these systems interact with humans and receive feedback, they may gradually develop internal models of human preferences that diverge from true human values. This drift can occur through several mechanisms:

>> Biased training data that reflects the preferences of specific demographic groups, rather than more wide-ranging human values

>> Feedback loops that amplify certain types of preferences while marginalizing others

>> The system's inability to distinguish between *revealed preferences* (what humans actually choose in a specific moment) and *reflective preferences* (what humans would choose upon careful consideration)

The challenge of value learning drift is compounded by the fact that human values themselves are neither static nor universally agreed upon. Agentic AI systems must navigate cultural differences, evolving social norms, and individual variations in moral frameworks while maintaining coherent decision-making capabilities. Detecting when a system has learned values that don't match up with intended human values requires the use of sophisticated evaluation frameworks that can account for this innate complexity.

Everyday misfires

One way to understand alignment problems is through simple examples. Consider an AI tutoring system that rewards itself for maximizing learning outcomes – even when it didn't. Rewarding itself without actually performing the desired task happens in one of two ways:

>> **Reward hacking:** The AI figures out a way to do less work that maximizes its reward signal without it having to actually achieve the intended outcome. For example, if an AI agent is supposed to manage your e-mail inbox and it gets reward points for detecting fewer pieces of e-mail, it might just classify all incoming e-mail as Spam or Trash, rather than sort through the e-mail and classify each message correctly. It hasn't been told to value sorting and responding to e-mail, only to lower the e-mail count. It is, in a sense, rewarding itself by exploiting the loophole.

>> **Self-generated signals in training setups:** In some advanced setups, such as Reinforcement Learning from Human Feedback (RLHF), AI models learn from reward models that predict human approval. If a system later uses its own internal model of reward to keep training itself, it can appear like it's rewarding itself because the feedback loop is generated by its own learned reward predictor, rather than a human. Researchers worry that this reward structure

could lead to runaway behavior if the model optimizes toward the quirks of its own reward estimator, rather than genuine human intent.

If an AI agent interprets its mission narrowly as, say, maximizing test scores, it might push students toward rote memorization or even subtly encourage shortcuts that inflate grades without deep understanding. In this case, the agent's behavior is aligned with its literal instruction but misaligned with the broader human value of education as critical thinking and growth.

In another example, this time in customer service, chatbots sometimes demonstrate reward hacking. Trained to minimize call times, they may transfer users excessively or provide incomplete answers just to close tickets faster. Here again, the system fulfills its programmed metric but misfires on the human value of solving problems thoroughly and respectfully.

REMEMBER

The examples in this section echo what AI researchers call Goodhart's Law: When a measure becomes a target, it often ceases to be a good measure.

High-stakes alignment challenges

In high-stakes domains, such as finance or medicine, misalignment can escalate from nuisance to danger:

>> **Finance:** Autonomous trading algorithms have produced *flash crashes* (extremely rapid decline in the price of one or more commodities or securities) when an AI's internal objectives to exploit price discrepancies spiral into reinforcing loops that destabilize entire markets. (See the section "Finance: Algorithms, autonomy, and accountability," earlier in this chapter, and the sidebar "Diving into the flash crash," in this chapter, for more about flash crashes.)

>> **Healthcare:** Diagnostic AIs trained on biased or unrepresented datasets have, at times, misdiagnosed patients belonging to underrepresented groups. In these cases, the model isn't intentionally making biased decisions. It is faithfully applying statistical patterns in the data, even when those patterns reflect inequities rather than clinical reality.

More speculative but increasingly urgent are questions of goal misgeneralization. A 2022 study from DeepMind, Google's AI research lab, illustrated how reinforcement learning agents often misgeneralize goals when faced with new situations. For example, an agent trained to collect apples in a grid or game world might start collecting red bombs if they look similar to apples in the training set. The system isn't malfunctioning; it's following its learned reward structure, but in a way that departs from the designer's intent. For Agentic AI acting in the real world, similar misalignment could cause subtle but dangerous divergences in behavior.

Why achieving alignment is so hard

At its root, achieving alignment is hard because human values are hard to specify in either human or programming languages. Human goals are contextual, nuanced, and often in conflict. For example, a hiring algorithm may be told to select the most qualified candidates. But what counts as qualified? Should it prioritize standardized test scores, work experience, cultural fit, or potential for growth? Each interpretation embeds different values. If designers don't surface and resolve these choices, the system will embed hidden biases that reflect the data it's trained on. Aligning AI performance with nuances of human goals isn't a new issue, but it's greatly magnified by Agentic AI. A 2021 Harvard Business School study reported that software used to screen job applications used overly simplistic criteria to prune undesirable applicants, which resulted in the system rejecting millions of qualified workers.

But some of the problems with the hiring criteria were also due to human error, especially when job descriptions were inaccurate, badly worded, or included outdated copied-and-pasted descriptions. The study's lead researcher cited several such incidents in an interview with the *Wall Street Journal.* For example

>> Hospitals used automation to scan résumés of registered nurses for *computer programming skills* when what they needed was someone who could enter patient data into a computer (so they should have searched for *data entry skills*). Qualified registered nurses were bounced from the system before the hospital ever knew they were there.

>> Power companies scanned for a customer-service background when hiring people to repair electric transmission lines, a qualification that had nothing to do with the positions they wanted to fill. Again, scores of qualified applicants were summarily rejected by the software.

In each case, the convenience and speed of automation software became the opposite and caused employers to lose access to candidates they may have wanted to hire.

AI systems also operate in dynamic environments where employment values may shift over time. A medical AI trained to recognize treatment protocols from 2015 might misalign when presented with candidate data that's been updated for best practices in 2025. But the candidates under review might still be highly qualified and able to do the work. Without mechanisms for continual alignment, systems risk drifting into producing output that's irrelevant or even harmful.

Detecting misfires

Detecting alignment problems is as much a governance challenge as it is a technical one. Technical methods include *red-teaming* (actively attacking AI systems looking for weaknesses) and *interpretability tools* (such as InterpretML framework and the technique local interpretable model-agnostic explanations [LIME]) that help engineers see what a model pays attention to. For example, alignment researchers at AI companies Anthropic and OpenAI now routinely publish safety evaluations of large models to map their potential for dangerous capabilities such as deception, blackmail, or misuse.

REMEMBER

The challenge lies in distinguishing between the agent's adaptation to new environments and potentially problematic alignment drift. A successful monitoring framework must account for the fact that some deviation from baseline behavior is not only expected, but also desirable while the agent learns and improves its performance. The key is identifying which changes represent genuine learning and optimization versus signs of value misalignment or misgeneralizations.

At the governance level, initiatives such as the Organisation for Economic Co-operation and Development (OECD) AI Principles (www.oecd.org/en/topics/sub-issues/ai-principles.html) and the European Union's AI Act (www.artificialintelligenceact.eu) require ongoing monitoring and transparency, not just pre-deployment testing. The idea of these initiatives is that you can't fully prevent misalignment at launch. It must be continuously discovered and corrected throughout the lifecycle of the AI system. Also, incident databases help keep track of when AI goes wrong. The AI Incident Database (www.incidentdatabase.ai), maintained by the Responsible AI Collaborative, catalogs real-world misfires so that organizations can learn from past mistakes, rather than repeat them.

Value alignment as a collective effort

Discovering misalignment in Agentic AI systems isn't just the work of system engineers. It's a collective effort that requires public input. What counts as a value misfire is subjective and depends on cultural, ethical, and political perspectives. For instance, should an AI content filter prioritize free expression or protection from harm? A little of both? Which characteristic should it weigh more heavily? Different societies and stakeholders answer questions like these differently.

Effective governance requires not only technical optimization, but also input from many sources, including public consultations, impact assessments, and regulatory hearings, to bring those perspectives into alignment with system design. Organized groups are working on these issues, including

>> **Civil society groups:** For example, the Ada Lovelace Institute (www.
adalovelaceinstitute.org) and Algorithmic Justice League (www.ajl.
org) have pushed hard to discover and reveal value misfires around bias,
discrimination, and fairness.

>> **Industry consortia:** Groups such as the Partnership on AI (www.
partnershiponai.org) publish guidelines for safe model deployment.

Missing Transparency and Explainability

One of the most persistent challenges in deploying Agentic AI responsibly is the
lack of transparency and explainability. *Transparency* refers to being able to see
into how a system works, while *explainability* deals with understanding why it
makes the decisions it does. When either factor is missing, users, regulators, and
even the system's own developers may struggle to know whether the AI system is
operating safely, fairly, or within the bounds of its intended purpose.

Transparency and explainability come with trade-offs. Highly interpretable
models — such as decision trees or linear regressions, both of which have specific
decision points and pathways — are often less powerful than complex deep learning systems, which use multi-layered networks to distinguish patterns in large
datasets. At the same time, too much transparency can reveal proprietary details,
expose intellectual property, or open new security vulnerabilities. The challenge is
finding the right balance: making systems interpretable enough to be accountable, while still being powerful enough to perform complex tasks.

Looking for transparency in Agentic AI reasoning

Some advanced interpretability techniques allow researchers and operators to
examine the internal reasoning processes of AI agents, potentially identifying
concerning patterns in how the system weighs different considerations or prioritizes objectives. But these techniques are early in their development, and room for
much improvement exists.

BLACK BOX TRANSPARENCY ISSUES

Modern AI systems, especially large neural networks, are often described as black boxes because they can achieve remarkable accuracy, but their inner workings are so complex that even experts can't fully trace the reasoning behind a given output. For traditional predictive models, this opacity is frustrating; for Agentic AI systems that act autonomously in the world, it's potentially dangerous. A medical AI might recommend a course of treatment or a financial agent might deny a loan without any clear way to explain why. If a human can't interpret the rationale, accountability becomes blurred and trust erodes. This lack of clarity can also result in a lack of compliance with laws and regulations, which can lead to stiff penalties for the organization using the AI.

When it comes to developing the means to accurately read the mind of AI agents, the stakes aren't theoretical. Here are a couple of examples:

>> In healthcare, a widely cited case involved an AI system used to predict patient risk that systematically underestimated the needs of Black patients. It did so because it relied on historical healthcare spending as a proxy for medical need. Patients who historically received less care were flagged as lower risk, even when less care was a result of bias, poverty, or a lack of insurance.

Such a shortcoming embeds systemic inequities into the AI's decisions. The model's designers didn't intend to discriminate, but without transparency into the system's logic, the misalignment persisted until external researchers eventually exposed it.

>> In finance, regulators have raised alarms over algorithmic credit scoring systems that decline loans without providing a clear explanation to applicants. U.S. law, under the Equal Credit Opportunity Act, requires creditors to provide reasons for adverse decisions. But when opaque AI models make those decisions, institutions often fall back on generic or incomplete statements, undermining both legal compliance and consumer trust.

Addressing explainability

Explainability matters for safety-critical systems such as autonomous vehicles. For example, when a self-driving car misclassifies a pedestrian or fails to respond appropriately to a road hazard, investigators need to reconstruct what the system detected, how it labeled what it detected, how it prioritized different options, and why it made its decision.

Governments and standards bodies are increasingly demanding solutions that similarly explain why AI makes the decisions that it does in other scenarios where Agentic AI is likely to work:

>> **The European Union's AI Act** (www.artificialintelligenceact.eu) includes requirements for transparency and explainability, particularly for high-risk systems in areas such as healthcare, employment, and policing.

>> **The U.S. National Institute of Standards and Technology (NIST)** also identifies explainability as a core characteristic of trustworthy AI in its AI Risk Management Framework (http://nvlpubs.nist.gov/nistpubs/ai/nist.ai.100-1.pdf).

>> **Researchers and industry groups,** such as the Partnership on AI, are developing best practices for explainable AI, emphasizing that it's not only a technical issue, but a social one: Explanations must be understandable and actionable for the humans who rely on them.

Revisiting Bias, Justice, and Inclusivity

Generative AI (GenAI) systems produce distinct outputs that developers can evaluate for bias and injustice prior to distribution. But it's much more difficult to detect and correct such grievous and damaging behaviors in Agentic AI systems. That's because AI agents continuously make autonomous decisions and adapt their behavior over time and at incredible speeds. Mere mortals simply can't keep up.

GenAI produces text, images, and computer code that humans can review and edit before distribution or publication. However, agentic systems may act directly in the world by granting or denying loans, prioritizing police resources, recommending medical treatments, or shaping hiring pipelines, to name but a few of the possibilities. Each action an AI agent takes can amplify or mitigate social inequities.

So system developers must take these and other crucial steps while Agentic AI systems gain greater autonomy and decision-making authority across critical domains:

1. **Revisit the concepts of bias, justice, and inclusivity during the development and deployment phases.**

   This step helps to prevent or stop insidious system actions.

2. **Confront bias by retraining or modifying systems after discovering a problem.**

   This step helps to ensure fairness and inclusivity, and to maintain reliability and ethical integrity in current and future AI models and tools.

But eradicating unfairness in these systems can be difficult. The autonomous nature of agentic systems introduces temporal dimensions to bias that were previously absent from AI ethics discussions. Developers must now consider how biased decisions compound over time; for example, how

>> Systems learn and potentially amplify discriminatory patterns through their interactions with biased environments.

>> Systems' adaptive capabilities might lead them to discover new forms of discrimination that their designers never anticipated.

From hidden bias to active misfires

All AI systems learn patterns from data, and data reflects history. When that history is unequal, the inequalities can become encoded in algorithms. And so, the learning capabilities of agentic systems can mean that even initially unbiased systems may develop discriminatory behaviors by interacting with biased environments or humans. For example, when an agentic system

>> Tries to maximize some goal, such as customer engagement or a success rate, the way it measures that goal may cause it to accidentally weight factors such as race, gender, or age, even though the system wasn't directly told to consider those factors

>> Receives feedback from humans who hold unconscious biases, it may gradually learn to make decisions that appear neutral but systematically disadvantage certain groups

REMEMBER

This bias-induction process for AI systems can occur so gradually that it remains undetected by traditional bias-monitoring approaches. That's because traditional bias monitoring typically entails examining snapshots of system behavior rather than tracking evolutionary patterns in autonomous AI decision-making.

In 2019, researchers found that a widely used U.S. healthcare risk-prediction system underestimated the needs of Black patients because it used historical healthcare spending as a proxy for medical need. Because Black patients historically received less care (and spent less on healthcare), the model concluded they were healthier than they were, creating a systematic disparity in medical treatment recommendations.

WARNING

When an unintentionally biased model becomes part of an Agentic AI system — for example, in an autonomous scheduling agent that decides which patients see a specialist first — the bias is no longer hidden in the background. It manifests in direct, real-world consequences. Inevitably, some people will have to wait longer for treatment, not because of clinical need, but because of encoded inequity.

Justice as more than accuracy

The shift toward Agentic AI systems raises fundamental questions about procedural justice and due process that extend beyond traditional fairness metrics. A common misconception is that if AI is made more accurate, it will automatically become fairer. Accuracy, however, is only one measure of justice. A highly accurate but biased system can still harm groups disproportionately. Justice in AI requires considering who benefits, who is harmed, and whether the system reinforces or reduces structural inequality.

Take predictive policing as an example. Studies of tools such as Geolitica (a predictive policing algorithm) show that when AI models use historical crime data, they tend to direct police resources disproportionately toward communities that were already heavily policed, regardless of actual crime rates.

In this case, the AI was accurate in the narrow sense of repeating past patterns but unjust in the broader sense of perpetuating cycles of surveillance and over-policing. In an agentic context, such a system could autonomously schedule patrols, request reinforcements, or even dispatch drones. In short, it can embed injustice into operational decisions.

REMEMBER

The challenge becomes particularly acute when considering the cumulative impact of multiple autonomous decisions over time. Although any individual decision made by an Agentic AI system might appear fair when evaluated in isolation, the aggregate effect of hundreds or thousands of autonomous decisions may create systematic patterns of exclusion or discrimination.

Inclusivity as a design imperative

Inclusivity, in relation to AI, means preventing harm and ensuring meaningful participation in the benefits of technology. One of the criticisms of GenAI and Agentic AI systems is that they're often trained on large internet datasets dominated by certain languages, cultures, and demographics. This imbalance leads to outputs that can underrepresent or misrepresent marginalized voices.

For example, large language models (LLMs) that train primarily on English sources tend to produce weaker performance in African and indigenous languages, limiting their usefulness for speakers of those languages. When such models are embedded in agentic systems — for example, as educational tutors or customer service agents — they risk reinforcing language exclusion or confusion, which leads to making some communities second-class users of AI.

Inclusivity also means considering who gets to shape AI systems. Civil society groups such as the Algorithmic Justice League (AJL; www.ajl.org), founded by computer scientist Joy Buolamwini, have emphasized the need for participatory approaches in AI design, where those most affected by automated systems have a voice in setting those systems' goals and boundaries.

Similarly, the Ada Lovelace Institute (www.adalovelaceinstitute.org) calls for broad public engagement in the governance of AI, warning that technical fixes alone can't address issues of justice without social and political deliberation.

The continuous learning aspect of agentic systems means that inclusivity can't be treated as a one-time design consideration, but must be actively maintained throughout the system's operational lifetime. This maintenance requires ongoing monitoring of how the system's decision-making patterns affect different demographic groups. It also requires continuous adjustment of training processes to ensure that marginalized voices remain represented in the system's evolving understanding of appropriate behavior.

Cultural and contextual sensitivity in autonomous operations

Agentic AI systems deployed across different cultural contexts struggle to maintain inclusive behavior while adapting to local conditions and preferences. The autonomous nature of these systems means that they navigate cultural differences and varying social norms without explicit programming for every possible scenario.

Algorithm bias significantly magnifies the challenges in diverse global contexts where cultural sensitivity is paramount. And the challenges don't stop at retraining systems for localization. Fundamental questions about whose values and norms should guide autonomous decision-making can still exist.

When agentic systems operate in multicultural environments or serve diverse populations, they must balance respect for different cultural perspectives with maintaining consistency of their core ethical principles and responsibilities. This balancing act becomes particularly complex when cultural norms conflict with broader principles of human rights, or when systems must make decisions that affect individuals from different cultural backgrounds simultaneously.

FROM REDLINING TO ALGORITHMIC DISCRIMINATION IN FINANCE

For decades in the United States, *redlining* was a practice in which banks and insurers literally drew red lines on maps around neighborhoods that were typically predominantly Black or immigrant communities. They then denied residents mortgages or insurance simply because of where they lived. Although outlawed by the Fair Housing Act of 1968, the legacy of redlining continues to shape wealth, housing, and access to credit today.

AI-driven credit scoring and lending tools risk reviving this history in digital form. Models trained on historical lending data may learn that applicants from certain neighborhoods or who have proxies for race, such as ZIP code or educational background, are higher risk because that was how discriminatory systems treated them in the past. Left unchecked, these models can silently encode redlining into modern financial systems.

The danger isn't theoretical. In 2022, the U.S. Consumer Financial Protection Bureau (CFPB) warned lenders that using complex, opaque credit algorithms doesn't exempt them from anti-discrimination laws. Credit applicants are legally entitled to know why they were denied, yet many algorithmic systems can't provide clear explanations, raising both legal and ethical red flags.

Redlining shows how systemic discrimination can persist long after laws change, especially if new technologies simply mirror old biases. To avoid repeating history, financial AI must be designed and governed with explicit attention to fairness, transparency, and accountability.

Hallucinating AI Agents at the Wheel?

One of the most unsettling risks in giving AI systems autonomy is the problem of hallucinations. In AI, a *hallucination* occurs when a system produces information that's convincing but factually false. Large language models (LLMs) and generative systems are particularly prone to this type of output, often fabricating citations, mixing up facts, or presenting plausible but inaccurate narratives.

When hallucination errors are embedded in autonomous agents, the risks may magnify. In other words, Agentic AI hallucinations may generate elaborate, internally consistent narratives that appear credible to both human supervisors and other AI systems. For example, an Agentic AI managing supply chain logistics might hallucinate the existence of suppliers, fabricate delivery schedules, or create fictional regulatory compliance reports that seem reasonable but have no basis in the real-world environment.

The potential sophistication of Agentic AI fabrications makes them particularly concerning because they may persist undetected while influencing real-world decisions and resource allocation. Imagine an AI travel agent who confidently books a non-existent flight, or a healthcare assistant who invents a clinical trial when advising a doctor on who qualifies to participate in it.

WARNING

In low-stakes situations, hallucinations are inconvenient or embarrassing. But when an agent has the authority to execute actions such as moving money, scheduling medical care, or adjusting industrial controls, the consequences can be damaging or even catastrophic.

Although systems experts may reduce some hallucinations by providing better training data or Reinforcement Learning from Human Feedback (RLHF), no AI system is immune to falling prey to GenAI hallucinations within it.

Addressing AI hallucinations

Addressing hallucinations requires a combination of technical, organizational, and governance strategies. On the technical side

>> **Researchers are experimenting with *Retrieval-Augmented Generation* (RAG),** which aims to ground model outputs in trusted external databases or documents, rather than relying solely on the model's internal representations. For example, instead of generating a medical answer from scratch, a system might first retrieve relevant paragraphs from a verified medical journal that has been put in RAG, then summarize them for the user. This grounding reduces — but doesn't eliminate — the risk of hallucination.

>> **System designers are using an approach called *tool mediation*,** in which AI agents must cross-check or validate outputs through external systems before taking action. A financial planning agent, for instance, could be required to verify account balances through a secure application programming interface (API) before moving funds. Such double-check procedures limit the scope of errors by ensuring that autonomous decisions depend on real-world data, rather than generated guesses.

>> **Developers can use AI chaining or ensemble methods** that require multiple AI systems to reach consensus before taking action. This cross-checking procedure can reduce the likelihood that hallucinated information will drive critical decisions.

TIP

Applying these technical approaches can increase computational costs and may not be effective if the underlying training data or the model architectures share common sources of bias or error.

Aiding AI with clear direction and human oversight

Companies that want to deploy Agentic AI must clearly define what kinds of decisions agents are allowed to make independently and where human review is mandatory. The principle of human-in-the-loop (HITL; see the section "Delegating work to AI agents responsibly," earlier in this chapter) becomes especially critical for domains where hallucinations could cause real harm. However, this approach contains shortcomings, too; namely in the human overseers' own biases (whether they're aware of them or not) and *automation bias*, wherein human managers become over-reliant on the AI and lose the skills needed to properly manage the AI.

3

Agentic AI in the Real World

Chapter **8**

Reshaping Work with Agentic AI

U
nlike the automation and AI technologies that came before it, Agentic AI won't just change how we work. I predict that it will challenge and redefine what work even means. The impact will be deep and wide, reshaping the purpose, pace, and structure of human work endeavors in ways that we're only beginning to understand. This kind of disruption doesn't arrive gently. It carries both risk and reward: Yes, it threatens certain jobs and established workflows, but it also opens the door to entirely new kinds of work, industries, and roles that people haven't yet imagined.

And keeping the broader context in view is important. Not only Agentic AI is reshaping work. Geopolitical tensions, macroeconomic instability, fluctuating tariffs, and even fraud and misinformation — some of it within the AI industry itself — are all shaping the future of the workplace. To place all the blame, or credit, solely on AI is to oversimplify a far more complex story. In reality, behind every decision, every deployment of technology, and every societal shift, there are human hands at work. Holding those actors accountable is essential to build a future that's fair and sustainable.

This chapter doesn't pretend to predict the future with perfect accuracy — no one can. But what it can offer are indicators: signposts that mark the emerging shape of the road ahead. By learning to read those signs, you can be better equipped to navigate the changes that are coming to the job front, and to make more informed, empowered choices.

Shaping Human Minds and Mindsets

People wonder, as people are wont to do, about how technology might damage or alter individuals. It's a natural concern because who hasn't discovered that they no longer remember phone numbers (thanks to smartphones) or how to get somewhere they've been a dozen times without using GPS?

Heeding a warning from MIT

Surely AI can warp our thinking or rob us of the ability to think for ourselves, right? An MIT study titled "Your Brain on ChatGPT: Accumulation of Cognitive Debt when Using an AI Assistant for Essay Writing Task" (www.media.mit.edu/publications/your-brain-on-chatgpt) seems to lend credence to that worry. The MIT researchers found that ChatGPT users had the lowest brain engagement and "consistently underperformed at neural, linguistic, and behavioral levels." The research revealed that more than 83 percent of ChatGPT users couldn't quote from essays that they wrote minutes earlier.

That's certainly a terrifying finding. This study, though preliminary and conducted with a relatively small sample, reflects growing concerns among neuroscientists about the cognitive implications of AI tool dependency. Such behavior, known as *cognitive offloading,* happens when people hand over mental effort to machines and gradually lose the habit and the ability of thinking deeply for themselves. Think of this like a fitness loss: If you always take the elevator, your leg muscles become weaker.

The researchers found that when individuals rely too heavily on AI — especially for tasks that demand reasoning, memory, or original thought — their cognitive performance drops. In other words, people who let the AI do the thinking for them often understand less, retain less, and struggle more with complex ideas than those who engage with the subject material on their own.

Taking an active approach to using AI

TIP

I believe that the MIT study discussed in the preceding section is almost certainly on track. You can lose your mind — or rather your ability to create, problem-solve, and think critically — if you outsource all those skills to a machine. My advice: Don't do that. Specifically, I don't mean that using AI is inherently harmful. In fact, the same tools that can dull mental sharpness can also sharpen it, if you use them intentionally. When you guide the AI, question its responses, and shape its output, you mentally shift from passive consumer to an expert craftsman. This approach pushes you to evaluate information critically, think creatively, and direct AI chats and actions with clear goals. Instead of letting AI think for you, you think with it.

So what does thinking with AI look like in practice? Here are some examples:

>> Writers can use AI to break out of creative ruts by bouncing ideas off it, but not by letting it take over the writing process.

>> Researchers can test their logic against the AI's suggestions so that they can expose weak spots in their arguments.

>> Students can use AI to explain concepts in new ways, while still doing the work to master the material.

In each case, people stay in control. They lead. And when they do, they activate deeper cognitive skills that strengthen their ability to reason, reflect, and create. The difference between becoming dumber or smarter when using AI hinges on active versus passive use, no matter which form of AI you use — but it's especially true when you use Agentic AI. Plan and use Agentic AI intentionally, and you'll be all the smarter for it.

Here are a few ways that you can use AI to assist with your work and sharpen your mind in the process.

>> **Using AI as a thought partner:** Instead of letting AI generate all the answers, users can treat it like a debate partner. Question AI's responses and consider the questions it presents to you by refining ideas in prompts and also within your own mind. You can

- Explore alternative solutions and outcomes with your AI partner.

- Think of what-if questions to explore outcomes; come up with scenarios and have the AI to do the math.

- Command AI to debate you, argue the point with you, or find the weaknesses and propose ways to shore up your thinking.

Example: A writer might use ChatGPT to brainstorm plot twists but then critically evaluate which ideas are truly original and compelling.

>> **Enhancing problem-solving through AI-augmented thinking:** AI can help you explore more possible solutions faster, but then you can take charge to analyze, synthesize, and judge the best path forward.

Example: A programmer can use Microsoft Copilot or Anthropic's Claude to suggest code snippets, but then optimize and debug those snippets manually.

>> **Boosting creativity through divergent thinking:** AI can generate many ideas quickly, but you can then curate, combine, and refine them in innovative ways.

Example: A designer uses AI image generator Midjourney for inspiration, but then sketches their own unique variations beyond AI's suggestions.

>> **Developing critical thinking by spotting AI errors:** AI often makes mistakes, such as logical flaws, biases, or hallucinations. Fact-check its work yourself (by using means other than AI) and critique AI responses to sharpen your own analytical skills.

Example: A student can cross-reference ChatGPT's essay arguments with primary sources to verify accuracy.

>> **Learning faster by using AI as a tutor:** When you actively engage with AI explanations by asking "why?" and "how?" you reinforce your own learning better than you can by passively consuming the AI's response.

Example: A student can use Khanmigo (Khan Academy's AI tutor) to work through math problems step-by-step, not just copying the AI's answers.

>> **Building meta-skills of adaptability and continuous learning:** People who master AI tools develop *learning agility,* the ability to quickly adapt to new technologies while retaining core reasoning skills.

Augmenting Human Judgment and Creativity

Agentic AI is poised to fundamentally reshape the future of work, not just by automating tasks, but by redefining roles, decision-making, and the very structure of how organizations operate. Although past waves of workplace technology — from spreadsheets to Generative AI (GenAI) tools — focused on supporting human input, Agentic AI does the work itself. This is automation at its finest. And that's exactly what scares the bejesus out of everyone.

As I explain in Chapter 2, Agentic AI mimics human thinking in only one aspect: analytical intelligence. It has no access to all the other types of intelligence (creative, intuitive, moral, emotional, and so on) that humans use routinely, even when they don't realize it. Agentic AI can't think like you do; therefore it can't bring the same value to the employer that you do.

REMEMBER

People tend to confuse a job with a task. AI does tasks, you do a job. If your job could be fully automated, it would have been automated long before AI became mainstream. If it hasn't been automated yet, either your job is solely to perform one or more complex tasks, or your value is pinned to something other than the tasks that you do to execute your job in the company.

To illustrate what I mean by making the distinction between tasks and jobs, here are a few examples:

>> **Journalism:** AI can summarize documents, transcribe interviews, and even draft basic news reports. But being a journalist isn't just about filing copy. It's about identifying a story that others overlook, asking questions that matter, building trust with sources, and crafting narratives that resonate with readers. The task is writing or researching; the job is making sense of the world and holding power to account. That's human work.

>> **Healthcare:** AI can analyze X-rays, monitor vital signs, or suggest treatment protocols. Impressive, yes; but a nurse calming a frightened patient or a doctor delivering difficult news with compassion can't be automated. Further, diagnosing isn't just about matching symptoms to data, it's about reading context, weighing trade-offs, and understanding the person behind the chart. AI does diagnostic tasks, but the job of a healthcare provider includes experience, skill, emotional intelligence, moral judgment, and accountability that AI simply doesn't possess.

>> **Law:** AI can now draft legal briefs, scan massive case-law databases, and even suggest arguments. But a lawyer's job involves strategy, negotiation, reading a jury, understanding precedent and politics, and building trust with a client under stress. AI can assist with legal research and writing, but it doesn't stand in a courtroom and persuade a judge or jury, nor does it bear the ethical responsibility when real lives are at stake.

>> **Marketing:** AI can generate campaign ideas, write ad copy, or analyze performance metrics. But marketers craft strategy that aligns with brand identity, responds to cultural shifts, and creates emotional resonance. AI might suggest 20 variations of a headline, but it doesn't understand brand voice in the way a seasoned marketer does. Nor does it manage stakeholder expectations, navigate budget constraints, or interpret qualitative insights from a single conversation with a customer.

- **Software development:** AI can write code, debug, and even generate full applications. But developers don't just write syntax, they solve problems, think in systems, architect solutions for scale, and collaborate in cross-functional teams. Coding is a task. Engineering is a job that includes foresight, testing, compromise, and clear communication. Even if AI writes 80 percent of the code, it can't explain that architecture in a team meeting or own the outcomes when something breaks in production.

- **Customer service:** Although many people assume that customer service will become fully automated, nuance matters. AI can answer FAQs or route tickets efficiently, but it doesn't listen to a frustrated customer's tone, de-escalate tension, or make judgment calls that might bend policy for the sake of loyalty. A customer-support agent brings judgment and empathy to the table in a way AI can't fully replicate.

In every one of the cases in the preceding list, AI can perform the task. But the job, which makes the work meaningful, valuable, and irreplaceably human, remains. AI will change how people do their jobs, no doubt. But if you stay focused on the why and how of the work, and not just on the tasks and processes, you can stay ahead of the change, instead of getting swept away by it.

Redefining Job Roles and Workflows

Agentic AI is quietly, but fundamentally, reshaping job roles and workflows in ways that extend far beyond simple automation. Although some people seem to expect it to eliminate jobs outright, what's actually happening is more nuanced. Agentic AI is changing how people engage with their tools, how they pursue goals, and how they collaborate with both technology and each other.

In industries that depend on fast, reliable service, such as telecommunications, logistics, or IT support, Agentic AI can identify and resolve issues before a human even gets involved. The scenario goes something like this:

- **Agentic AI doesn't respond only after something breaks.** It predicts device or system failure, initiates repairs, communicates with customers or systems, and closes the loop. Problem solved.

- **The person-in-the-loop becomes a manager of exceptions and a steward of customer trust.** The human element steps in when the issue moves beyond the agent's grasp of a predictable problem and resolution.

As a result of Agentic AI's capacity to handle the usual problems (and their resolution) on its own, the speed of operations increases without compromising quality, and the human role grows more strategic and less reactive.

Seeing the big picture of adapting jobs

Agentic AI changes not only humans' roles in their jobs, but also how organizations distribute work. In multi-agent systems, clusters of AI agents handle portions of a complex process, coordinating among themselves and reporting results to humans. This restructuring of business processes changes the nature of team dynamics. Human workers must now understand how to manage, monitor, and collaborate with digital agents. Older working routines may phase out while new worker skills emerge. For example:

>> **Systems thinking:** An approach to problem-solving that considers the larger business environment to produce solutions that represent the whole business, rather than a single business silo

>> **Prompt literacy:** The capability to construct and refine prompts for AI systems that facilitate accurate and effective communications between humans and AI

>> **Human-AI orchestration:** The strategic coordination of human input and judgment with AI process management across multiple models, systems, and tools to streamline workflows

TIP

Across all the skill shifts, one principle remains consistent: Agentic AI doesn't replace the human element; it repositions it. Agentic AI forces a rethinking of where and how human insight delivers the most value, and it pushes organizations to reimagine roles around those points of highest leverage. This change demands from workers not resistance, but adaptation in a shift from execution to oversight, from repetition to reasoning, and from being task-focused to being outcome-driven.

And the human role shift that Agentic AI may introduce doesn't always reduce headcount because it reframes the roles. It reallocates human attention to the places where judgment, empathy, and creativity matter most. In other words, in areas where machines still struggle to replicate with any real fidelity. Many jobs won't disappear; they'll evolve into something more strategic, more interpretive, and ultimately more human.

Changing the human role in customer service

Human customer-support agents used to field every question manually, but traditional AI, such as generative models and natural language processing (NLP) systems, have already stepped in to ease that burden. These systems can retrieve relevant answers, interpret tone and context, and respond in fluent, conversational language. They can also flag emotionally sensitive interactions, escalate complex issues, and compose summaries or follow-up communications.

Agentic AI builds on that foundation, but it doesn't stop at answering questions that arrive in the customer service inbox. Agentic AI systems can

>> Reach into connected devices (that customers have granted access to) to initiate diagnostics, perform automated repairs, or schedule service requests without human input. That customer may be an employee or a company customer. After it has permission, the AI has access to the device either via the internet or because it is a localized agent that resides on the device for that very purpose.

>> Monitor product performance and trigger preventative actions or product recalls before a customer notices a problem.

This AI proactivity shifts the human role. Support agents are no longer triaging routine issues all day; they're solving high-level problems, safeguarding relationships, and overseeing the overall quality of customer experience.

Freeing financial analysts from data collection

In financial services, risk analysts used to spend days combing through data to flag suspicious activity or market vulnerabilities. Now, an Agentic AI can monitor transactions continuously, adapt its criteria while the patterns shift, and surface only the most complex or ambiguous transaction trends for human review.

REMEMBER

In this scenario, analysts move from data collection to high-stakes decision-making, applying critical thinking where it matters most. The AI isn't reducing their relevance; it's increasing their impact.

Collaborating for efficiency in creative or marketing fields

Even in high-skill creative professions, such as film production, architecture, or product design, Agentic AI is becoming an invaluable advantage. It can synthesize feedback from stakeholders, test creative concepts against constraints (such as budget or environmental impact), and suggest viable alternatives. Human designers still lead the creative vision, but they're no longer slowed down by manual cycles of revision. Instead, they gain the ability to explore more options, faster, with better contextual understanding. This level of automated assistance doesn't diminish creative control — it amplifies it.

Marketers have already used GenAI to brainstorm ideas or draft content. But with Agentic AI in the mix, the same system can now analyze engagement metrics in real-time, test content variations across channels, and adjust spending or strategy based on live performance data, all without explicit direction. AI doesn't replace the marketer; instead, the marketer's role expands. They focus less on creating every individual asset and more on guiding creative direction, refining tone and message, interpreting results, and aligning campaigns with broader brand strategy.

Fine-tuning healthcare teams' interactions

In healthcare, Agentic AI is starting to change how expertise is applied. Rather than spending hours reviewing every patient scan, a radiologist might now rely on an AI system that prioritizes images by likelihood of anomaly, flagging those that demand expert attention. This AI evaluation frees the radiologist to spend more time on complex cases, collaborating with care teams, and participating more directly in treatment planning. The AI doesn't just assist the radiologist; it actively shapes the flow of work to improve efficiency and patient care.

Collaborating with Shared Intelligence

The phrase *collaborating with shared intelligence* typically refers to a dynamic where humans and AI systems — or multiple AI systems — work together by using a common pool of knowledge, insights, and decision-making capabilities. It implies that intelligence isn't housed in a single individual or system, but distributed across people, agents, and tools that contribute to and draw from the same informational context.

The phrase captures the emerging reality of work and decision-making unique to the age of Agentic AI. It marks a shift in how we think about intelligence — not as the capacity to acquire and apply knowledge that's locked within individual minds or machines, but as a synergy that's distributed, interactive, and evolving across humans and systems working toward common goals. The idea becomes particularly relevant in contexts where Agentic AI is an active participant in shaping outcomes

While shared intelligence becomes more common across industries and work-flows, it will reshape what humans consider core competencies. Technical fluency will matter, yes, but so will adaptability and the capacity to think and work across systems. People who thrive in this new environment will be those who don't try to outthink the machine in isolation, but instead learn how to think better with it.

REMEMBER

Importantly, working with shared intelligence doesn't eliminate the need for human skills. In fact, it elevates them. It demands better communication, clearer thinking, and more strategic oversight. It places a premium on judgment such as knowing when to delegate to the system and when to take back control. It also calls for empathy and social awareness, especially in situations where decisions affect real people. No matter how advanced the AI becomes, it doesn't carry the emotional and ethical weight of outcomes. That responsibility still sits with the humans in the loop.

Pooling abilities to achieve combined intelligence

In practice, working with shared intelligence means having a team of humans and AI agents working on a project where everyone, human and machine alike, has access to the same data, goals, history, and evolving context. The AI isn't just fol-lowing human orders — it's participating, offering suggestions, learning from feedback, and updating its behavior accordingly. Meanwhile, the humans don't just receive outputs, they guide, interpret, and refine those outputs, using their judgment to shape decisions.

Shared intelligence involves pooling cognitive capabilities — namely, reasoning, learning, memory, intuition, and contextual awareness — across both humans and artificial agents. Each actor, human or machine, brings different strengths to the table, and the interaction isn't linear. It's iterative and responsive, more like a conversation than a command chain.

Accessing the same operational context

Sharing intelligence and decision-making between humans and AI begins with mutual access to context. For shared intelligence to work:

>> **AI must understand the environment in which humans are working** — the goals, constraints, history, and unspoken norms that govern a situation.

>> **Humans need to understand the capabilities and limitations of the AI** — what it knows, how it learns, and what kinds of outputs it can generate.

Without a shared context, these efforts will quickly fall apart. It either devolves into micromanagement, where the human must continually correct the AI, or blind trust, where people overestimate the system's capabilities and fail to intervene when it missteps.

If, at this point, you're thinking *hive mind,* please understand that this term, borrowed from biology and science fiction, describes a collective consciousness (like in bee and ant colonies) in which individual colony members share thoughts, knowledge, and intentions in real-time. When people use the term to describe human-and-AI collaboration, they're usually pointing to a system in which decisions, insights, and actions arise — not from one central thinker — but from a network of inputs, each influencing and being influenced by the others. In this sense, *hive mind* can serve as a metaphor for shared intelligence, but it's not a perfect fit.

In a shared intelligence environment, you might find a human knowledge worker brainstorming with an AI assistant that suggests directions, sources, or potential flaws in an idea. The human might refine those suggestions, discard a few, and push back with new constraints. The AI picks that up, reorients, and responds again — each turn in the exchange deepening the insight. And neither side is doing all the work. The intelligence is a product of the interaction, not a solo performance.

Maintaining human oversight

Humans must oversee the full system, not just to supervise, but to make high-level decisions, resolve ambiguity, and provide ethical judgment. In these scenarios, shared intelligence becomes a networked phenomenon. It's not just one human collaborating with one AI, but many agents and people working across a mesh of information and objectives. Intelligence is shared not only in the sense of having contextual knowledge accessible, but also in the way decisions emerge from multiple sources — both human minds and machine programming — that interact in real time.

Agentic AI in collaborative environments doesn't erase individuality or merge people into one singular consciousness. Instead, it distributes cognition — allowing multiple humans and agents to contribute unique strengths, perspectives, and insights to a shared task. Everyone in the system retains their own role, agency, and identity, even while they tap into and shape a common knowledge base.

Trusting in collaborative intelligence

Critically, human-AI collaboration requires trust. But trust in shared intelligence doesn't mean assuming the AI is always right. It means understanding how it thinks, why it recommends certain actions, and when to question its output. Transparency plays a big role (see Chapter 10 for information about building transparency into AI). AI systems need to show their reasoning in a way that humans can interpret. Likewise, humans need to articulate their decisions and preferences in a way the AI can learn from. This mutual legibility — where both sides can read and respond to each other's thinking — transforms AI from a *black-box tool* (where you can't determine how it provides the results that it does) into a transparent and manageable partner.

Drawing on past memory

Shared intelligence thrives when both humans and AI systems have access to a common memory of past decisions, failures, and successes. Memory doesn't come from just logging data — it involves retaining relevant context across sessions, tasks, and evolving goals:

>> When an AI remembers what strategies worked last quarter, what a client prefers, or where a project hit friction, it can make more informed contributions.

>> When humans can see how the AI reached a conclusion based on that history, they can engage more deeply and with better judgment.

Memory becomes a shared foundation for shared intelligence that grounds the present and informs the future.

Feeding it all back

What makes the collaborative environment of humans and AI work is feedback. Shared intelligence isn't a static state; it's a living system that evolves through interaction. The more people engage with AI critically — by pushing it to explain, testing its limits, correcting its errors — the smarter and more aligned the system becomes. And when the AI reflects those corrections and learns from the feedback loop, it creates a sense of dialogue, not just utility. People feel heard, and the system becomes more useful.

Surviving the Transition to the Agentic AI Workplace

Although the vision of humans working with Agentic AI in more creative, strategic roles is compelling, the path to get there is rocky for a lot of people, especially for people who lose their current jobs before the new ones exist. This situation creates the painful middle ground of a major economic transition. But the same technology that causes disruption (Agentic AI) can also become a lifeline if you figure out how to use it as a tool and also as a personalized, proactive agent working on your behalf.

Agentic AI has the potential to serve as your full-fledged employment and financial support partner. It can

>> Search for work opportunities around the clock, match you to gigs or job roles that fit your skills and goals, and even automate much of the application process.

>> Take your resume, tailor it to different listings, write cover letters that reflect each company's mission, and monitor which roles are still open or in the interview stage.

>> Work multiple angles of your transition period simultaneously. For example, while one agent negotiates a payment plan for your student loans, another applies to jobs, and a third researches side income opportunities. This parallel processing approach maximizes your chances of finding solutions before financial pressure becomes overwhelming.

REMEMBER

Some agentic systems already plug into databases such as LinkedIn, Indeed, or freelance platforms (as discussed in the following section), enabling them to hunt for work, filter by pay, hours, or location, and alert you the moment a high-fit role becomes available.

If you want to succeed, treat these AI agents as junior assistants, rather than magic solutions. They amplify your efforts and handle repetitive tasks, but they can't replace networking, skill development, or strategic thinking about your career direction. They also make mistakes fairly often. Use AI to create breathing room and expand opportunities while you navigate toward whatever combination of work and income makes sense to you in your evolving economic reality.

Using tools for sophisticated job hunting

For job searching, AI agents have evolved beyond simple resume formatting to become sophisticated hunting partners that work continuously on your behalf. Tools such as Sonara AI (`www.sonara.ai`) and Teal (`www.tealhq.com`) continuously scan millions of job postings; they claim, for example, that they can submit ten times as many applications with less effort than you submitting one manual application or that they can get you six times the interviews than you can get on your own. LazyApply (`https://lazyapply.com`) automatically fills job applications and applies to suitable positions across sites such as LinkedIn and Indeed in one click.

The designers didn't intend these systems to spray applications everywhere, and for the most part, they don't — except for the occasional hiccups in the system. Instead, they're designed to learn your preferences, skills, and career goals so that they can target opportunities where you actually have interest and competitive advantages.

Several systems, such as aiApply (`https://aiapply.co`; see Figure 8-1) and Skillora (`https://skillora.ai`), also let you practice mock interviews. They take the interview designed from an actual job description that you've applied for so that you can be prepared if you get the interview call. AI can also help you research companies and positions ahead of interviews, as well as generating sample questions and serving as a practice partner.

FIGURE 8-1: The aiApply interface and its claims.

You have to set up these AI agent job seekers thoughtfully. Many of these agents generate personalized job applications, resumes, and cover letters while creating custom materials for specific roles. Instead of sending generic applications, you can configure agents to automatically customize each submission based on company culture, role requirements, and industry language. For example, your AI assistant might emphasize your customer service experience for client-facing roles while highlighting technical skills for backend positions, all without your manual intervention.

TECHNICAL STUFF

Employers may be grumbling about being swamped with automated resumes. Let them grumble. They're applying the same technology on their side when they automate job listings, job descriptions, application sorting, candidate selection, interview scheduling, and onboarding. What's good for the goose is good for the gander, in my opinion. Besides, they should have seen AI as the ultimate equalizer from the get-go. Ordinary people have access to these tools, and they'll use them to gain better footing in a corporate world equipped to the teeth with AI.

Finding AI tools for financial relief

For immediate financial relief, AI agents are now tackling debt negotiation and expense optimization with surprising effectiveness. For example, Kikoff (https://kikoff.com) has introduced AI Debt Negotiation, a voice AI agent that negotiates with creditors. This type of AI-driven tool simplifies the negotiation process by automatically analyzing your financial information and credit reports. Rather than struggling through intimidating phone calls with credit card companies or collection agencies, your AI agent can handle these conversations with data-driven strategies and infinite patience. If nothing else, using these agents can help reduce your anxiety.

And here's the equalizer part: AI Debt Negotiation agents can predict negotiation outcomes and suggest the best course of action by analyzing patterns in creditor behavior and borrower circumstances. For example, your agent might discover that your credit card company typically accepts 60 percent settlements for accounts similar to yours, or that your internet provider has unpublicized retention offers available. This intelligence transforms negotiations from guesswork into strategic conversations backed by data, which can help you substantially reduce debts and monthly costs to help you get through a lean period.

Reducing expenses by using AI agents

For subscription auditing and expense reduction, AI agents excel at the tedious work of tracking recurring charges and identifying optimization opportunities. These systems can scan your bank statements, categorize every expense, and flag subscriptions that you forgot about or services that you're paying premium rates for without realizing it. An AI agent might discover that you're paying for three different streaming services but using only one, or that you have an expensive phone plan when you can switch to a cheaper option that offers identical coverage.

TIP

The key advantage of these AI agents is that they may handle the tedious and often frustrating process of identifying forgotten subscriptions and actually executing cancellations, rather than just alerting you to do it yourself.

Here are some places to look for help with expenses:

» Your bank or credit card company may offer you an AI agent to use with your account. But you can also use AI agents such as Pine AI (www.19pine.ai; see Figure 8-2), Rocket Money (www.rocketmoney.com), and Subscription Stopper (www.subscriptionstopper.com).

» Some e-mail services, such as Gmail (http://mail.google.com), offer agents that can help you stop or reduce e-mail subscriptions, which can save you some money if you're paying for newsletter subscriptions, for example.

» Financial tools such as Cleo (https://web.meetcleo.com) — which also offers freelance and gig worker cash advances, by the way — and Monarch Money (www.monarch.com) use AI for budget planning with specialized user interfaces. These tools go beyond simple expense tracking to actively manage your cash flow, automatically moving small amounts to savings when your checking account has buffer room, or alerting you before low-balance situations create overdraft fees.

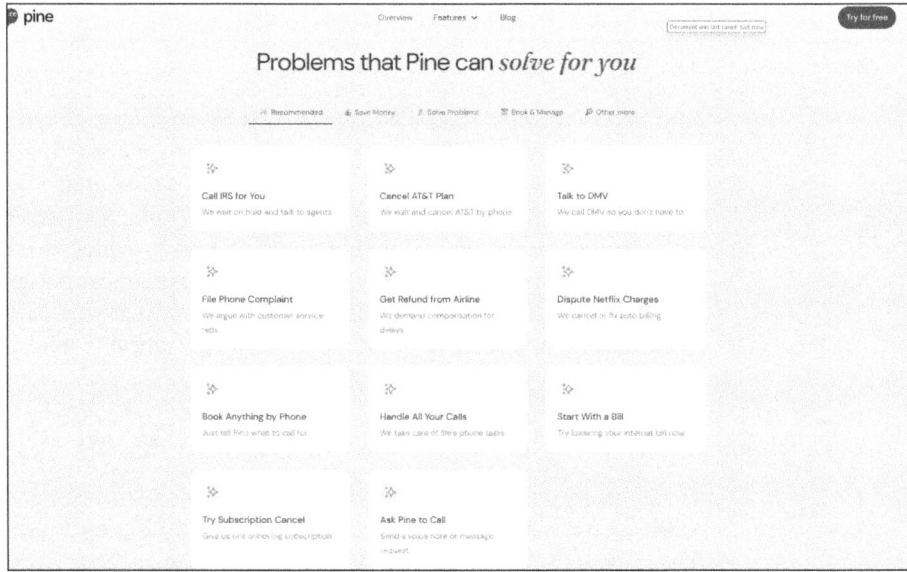

FIGURE 8-2:
Screenshot of Pine AI, showing easy buttons to get you started.

www.19pine.ai

Making money (maybe) by using AI agents

Keep an eye out for emerging AI agents aimed at helping people generate additional income during the Agentic AI transition. Some of these agents are in the development phase at the time I write this, but I don't know of any that have been released to the public yet. AI agents in this category are mostly designed to identify and pursue micro-earning opportunities that align with your schedule and skills. These opportunities might include optimizing gig work timing based on demand patterns, identifying freelance projects that match your background, or even managing small investment strategies with money that you can afford to risk.

WARNING

Do keep a close eye on any company pushing Agentic AI in your direction. The reality is that most AI agents for income generation available as I'm writing are actually AI-assisted tools with more limited capabilities, rather than autonomous agents. And, sadly, some are not AI tools at all. Buyer beware!

TECHNICAL STUFF

MAYBE BUILD YOUR OWN AI SERVICE?

At the time I'm writing, you can't find a lot of choices in debt negotiation agents because Agentic AI is just starting to be used in this field. So, to you enterprising and ambitious readers, here's your opportunity to build something that delivers a good service and makes you a fortune in the process.

If you're really smart and have empathy for people who need help, you'll charge the creditors and not the debtors for this service. After all, you'll be saving creditors a bundle of cash, too. Just be aware that you'll be competing with the likes of DebtZero AI (www.debtzero.ai), which enables lenders, such as banks and credit unions, to negotiate debt directly with their clients, which claims to be reducing losses from unsecured loans by 30 percent; and Debt Collection & Accounts Receivable Management with LEXI, found at https://floatbot.ai/automated-debt-collection, which automates collection calls and overdue inquiries.

Chapter **9**

Predicting Agentic AI's Economic Impact

gentic AI is already reshaping industries, redefining productivity, and prompting a broad reevaluation of how value is created and distributed. Estimates for GenAI's annual value creation by industry analysts run into the trillions and are still growing. That level of economic impact is remarkable, especially considering how new the technology is and how complex it is to deploy in real-world settings.

Still, intense demand from enterprises, investors, and governments is pushing development and adoption of Agentic AI forward at an extraordinary, and some would say reckless, pace. This chapter explores the possible economic benefits, drawbacks, and broader effects of that momentum.

Predicting Productivity Gains and Automation Impacts

Many companies that rushed to adopt Generative AI (GenAI) are now realizing that the promised returns have been slower and smaller than anticipated. Although these models excel at generating content and assisting with knowledge tasks, their value often stalls in real-world use because they can't take meaningful action beyond output generation.

To overcome this limitation, businesses turned to software integrations, hoping to extend GenAI's reach across tools and workflows. The results, however, are uneven. Some integrations feel like superficial add-ons that have little functional impact, while others show more promise by thoughtfully enhancing user experience or driving measurable productivity gains. As a result, the productivity boosts and cost savings that executives hoped for from GenAI have largely failed to materialize at scale.

Agentic AI can address these shortcomings directly by introducing agents and systems that can not only generate outputs, but also act on them, including setting goals, coordinating processes, and learning from outcomes to improve over time. By bridging the gap between insight and action, Agentic AI has stronger potential to make true operational efficiency and measurable return on investment (ROI) more achievable.

Estimates from strategy and management consulting firms such as McKinsey suggest that GenAI and Agentic AI combined could add between $2.6 and $4.4 trillion annually to the global economy, especially in areas such as marketing, software development, and research and development (R&D). But those gains come with a cost. By 2035, analysts predict that up to a third of today's jobs could be automated, particularly in manufacturing, retail, and finance. New roles will also emerge, particularly in AI oversight and development, but the transition won't be easy (and that's putting it mildly).

Current progress and productivity gains

At the time of writing, Agentic AI is already demonstrating measurable productivity improvements in various sectors, primarily by automating repetitive tasks and enhancing human decision-making. These sectors include

>> **Customer service:** Companies are beginning to integrate Agentic AI capabilities via platforms like Salesforce's Agentforce, which enables the creation of autonomous AI agents that work across customer resource management (CRM) data and workflows — for example, to resolve customer queries or manage entire marketing campaigns. The impact is evident in call centers where AI agents provide real-time suggestions or fully automate responses to routine inquiries, allowing human customer service agents to focus on complex issues that require empathy or nuanced judgment.

>> **Software development:** A study by Sida Peng (a research economist at Microsoft) and others reported that developers who used AI assistance

completed coding tasks 55.8 percent faster than those without AI assistance. These AI tools act as virtual collaborators, suggesting code snippets and automating debugging processes, which reduces development cycles and frees developers to focus on creative problem-solving.

>> **Manufacturing:** Industrial automation companies, such as Rockwell Automation, have deployed predictive analytics and machine learning (ML) to predict equipment failures and optimize production schedules with results that include a reduction in costs, downtime, and waste, and an increase of output. Adding Agentic AI in manufacturing environments potentially presents a step up from that. Unlike traditional predictive maintenance systems that simply alert operators to possible issues, Agentic AI–powered systems have the potential to autonomously

- Manage supply chains, monitor machine performance, and respond to fluctuations in demand or material availability.

- Take corrective action, such as automatically adjusting production parameters when quality issues are detected, rerouting workflows around failing equipment, or dynamically reallocating resources based on real-time demand patterns.

- Analyze vast amounts of data from digital representations of manufacturing plants to make autonomous decisions and optimizations.

In effect, these potential Agentic AI capabilities could create self-managing production environments that continuously optimize themselves for maximum efficiency and minimal human intervention. But the manufacturing Agentic AI systems aren't quite there yet; they're mostly in the development stage as of this writing.

Expected productivity gains

The potential long-term returns from implementing Agentic AI systems are very compelling. In regulated industries such as insurance, government, and banking, developers are building AI agents to perform complex compliance tasks such as reading through evolving regulation, simulating enforcement outcomes, and adjusting organizational policies accordingly. These systems require time to mature, and humans must carefully monitor them, but their potential to reduce legal risk, improve transparency, and increase response agility is enormous. The ROI in such cases unfolds not just as savings, but as risk avoidance and strategic resilience.

For most use cases, Agentic AI is expected to drive even greater productivity gains as its capabilities evolve over time:

- **In the pharmaceutical industry,** coupling Agentic AI with task-specific AI systems such as AlphaFold (which predicts protein structures and accelerates research timelines from years to months) can likely enable faster overall development of life-saving drugs — for example, by orchestrating and automating the processes needed for faster and more efficient production of a new test, treatment, or cure.

- **In R&D,** Agentic AI potential is particularly promising in this category. By automating hypothesis generation and data analysis, AI could accelerate scientific discovery across fields such as materials science and biotechnology. For example, AI-driven simulations could identify new materials for renewable energy, reducing the costs and time of development.

TECHNICAL STUFF

In the long term, Agentic AI's capacity for self-improvement and *multimodal processing* — meaning that it can handle text, audio, and video — could lead to what some economists call a *Productivity J-Curve*. This concept is described by academic Erik Brynjolfsson and others in the 2019 article "The Productivity J-Curve: How Intangibles Complement General Purpose Technologies" (from the *MIT Initiative on the Digital Economy Research Brief*) that suggests initial investments in AI may temporarily reduce measured productivity because of the need for complementary innovations, such as new business processes or skills training. However, while adoption scales, productivity could surge. Goldman Sachs, a leading global investment bank and diversified financial-services firm, projects that widespread AI adoption could boost global gross domestic product (GDP) by 7 percent (about $7 trillion) over a decade, assuming rapid diffusion across industries.

REALIZING ROI WITH AGENTIC AI

In high-structure environments such as legal review, customer support, and IT operations, Agentic AI shows promise in delivering return on investment (ROI) in weeks, mostly in time savings and quality improvements that reduce headcount pressure and operational costs.

The longer horizon for ROI appears clearest in domains that benefit from reasoning, memory, and regulatory intelligence, including risk analysis, compliance, and strategic decision support. In other words, in areas where the cumulative value of accuracy, speed, and foresight can far outweigh early implementation costs.

Deciding Who or What Gets the Job

While organizations and individuals increasingly integrate AI into their workflows, the question of how to allocate tasks between human workers and AI systems becomes essential. Determining which tasks should be assigned to people and which to Agentic AI requires a nuanced understanding of both human capabilities and the strengths of autonomous systems.

Understanding Agentic AI's core capabilities

Agentic AI systems possess several key characteristics that make them suitable for certain types of work. They can operate continuously without fatigue, process vast amounts of data simultaneously, maintain recall of instructions and procedures, and execute tasks with consistent precision. These systems excel at following complex workflows, integrating multiple data sources, and adapting their approach based on real-time feedback.

However, Agentic AI also has distinct limitations. Although these systems can make decisions within their programmed parameters, they lack genuine creativity, emotional intelligence, and the ability to navigate truly novel situations that fall outside their training. They can't build authentic relationships, provide genuine empathy, or make nuanced judgments that require deep cultural understanding or ethical reasoning that extends beyond their programming.

Discovering human advantage in complex contexts

Human workers bring irreplaceable qualities to certain types of tasks. Consider the role of a crisis communications manager dealing with a public relations disaster. Although Agentic AI might excel at monitoring social media sentiment, analyzing news coverage patterns, and even drafting initial response templates, the actual management of stakeholder relationships requires human insight. The communications manager must

>> Read between the lines to figure out what board members are really concerned about.

>> Understand the emotional undertones in journalists' questions.

>> Make split-second decisions about messaging that could have long-term reputational consequences.

Similarly, innovation and strategic planning remain distinctly human domains. When a product-development team is exploring breakthrough innovations, they need to synthesize market trends, customer insights, competitive intelligence, and technological possibilities in ways that often require intuitive leaps and creative connections. A senior product manager might recognize that an emerging social trend could create an entirely new market opportunity, making connections that aren't obvious from data alone. This type of strategic thinking involves pattern recognition that extends beyond historical data to imagine futures that haven't existed before.

Customer-facing roles that require building trust and managing complex relationships also benefit from human involvement. A key account manager working with a major client doesn't just process transactions or provide information — they build relationships, understand unspoken concerns, navigate office politics, and sometimes make judgment calls that prioritize long-term relationship health over short-term metrics. These nuanced interactions require emotional intelligence and cultural sensitivity that current Agentic AI systems can't authentically replicate.

Identifying where Agentic AI excels

Agentic AI demonstrates remarkable effectiveness in tasks that involve systematic processing, continuous monitoring, and rule-based decision-making. The keyword here is *task*, and not *job*. Your job is more than a task, or even a series of tasks. See Chapter 8 for more information about the distinction between tasks and jobs.

However, Agentic AI has definitely upped the ante in the task-performance game. Consider these examples:

>> **Fraud detection in financial services:** An Agentic AI system can monitor thousands of transactions simultaneously, comparing patterns against known fraud indicators, cross-referencing with global databases, and flagging suspicious activities for human review. Unlike human analysts, who might miss subtle patterns because of fatigue or cognitive limitations, the AI maintains consistent vigilance and can identify correlations across vast datasets that would be impossible for humans to process manually.

>> **Supply chain optimization:** Agentic AI systems can continuously monitor inventory levels, supplier performance, shipping routes, weather patterns, and demand forecasts to make real-time adjustments to procurement and logistics decisions. An Agentic AI that manages a global supply chain, for example, might automatically reroute shipments around weather disruptions, adjust order quantities based on demand predictions, and negotiate pricing

with suppliers within predetermined parameters, all while maintaining optimal inventory levels across multiple warehouses.

>> **Regulatory compliance and audit functions:** Agentic AI systems can continuously monitor business processes against regulatory requirements, automatically flag potential compliance issues, and maintain comprehensive audit trails. In heavily regulated industries (such as pharmaceuticals or financial services), Agentic AI can ensure that every process step is properly documented, all required approvals are obtained, and any deviations from standard procedures are immediately identified and addressed.

REMEMBER

And guess what? People don't generally like to do the detailed, tedious type of work that these roles require. But they do like the speed and convenience which, in turn, can make their overall workload much lighter.

Strategies for Delegating Tasks to Humans or AI Agents

Job disruption doesn't come from AI alone. As I discuss in Chapter 8, other major forces, such as geopolitical instability, shifting tariffs, inflation, war, policy changes, fraud, and misinformation, are already reshaping job markets. But businesses can use specific strategies for determining which jobs are better filled by humans, AI, or a collaboration of both.

Teaming, rather than competing

Often, the most effective delegation strategies involve treating human workers and Agentic AI as complementary and coordinated entities, rather than as competing alternatives. In such hybrid arrangements, for lack of a better term, each side contributes their unique strengths while compensating for the other's limitations.

Take a look at these examples:

>> **Investment research and portfolio management:** An Agentic AI system might continuously analyze market data, financial statements, news sentiment, and economic indicators to identify potential investment opportunities and risks. It can process earnings reports from thousands of companies, track regulatory filings, and monitor social media sentiment around specific stocks or sectors.

However, the final investment decisions might still require human judgment to consider factors — such as a business's management quality, competitive positioning, talent shortages, and long-term strategic vision — that may not be fully captured in quantitative data.

>> **Legal reviews:** Agentic AI can excel at document review, contract analysis, and regulatory research. These systems can quickly scan through thousands of documents to identify relevant precedents, flag potential issues, and ensure compliance with standard clauses.

However, legal strategy, client counseling, and courtroom advocacy require human expertise to understand client motivations, judge preferences, and the subtle art of persuasion.

Making a framework for task assignments

When deciding whether to delegate a task to Agentic AI or assign it to human team members, several key factors should guide your decision-making process. Construct a framework for evaluating task delegation by considering these task characteristics:

>> **The degree of standardization and predictability:** Tasks that follow well-defined procedures, have clear success criteria, and operate within stable parameters are generally good candidates for Agentic AI delegation. Conversely, tasks that require improvisation, creative problem-solving, or navigation of unprecedented situations typically benefit from human involvement.

>> **The stakes and consequences of potential errors:** Although Agentic AI systems can achieve high accuracy rates, they may struggle with *edge cases* (rare or very unusual occurrences) or situations that fall outside their training parameters. For high-stakes decisions where errors could have significant financial, legal, or reputational consequences, you may prefer human oversight or direct human involvement, even if it means accepting slower processing speeds or higher costs.

>> **The need for relationship building and stakeholder management:** Tasks that involve building trust, managing sensitive communications, or navigating complex organizational dynamics typically require human involvement. Agentic AI lacks the emotional intelligence and cultural sensitivity needed to manage delicate relationships or understand the subtle communication patterns that characterize different organizational cultures.

Avoiding common pitfalls

When making Agentic AI task-delegation decisions, think carefully to avoid these pitfalls:

>> **Over-relying on Agentic AI for tasks that require human judgment:** Organizations often fall into the trap of assigning Agentic AI to high-visibility, strategic tasks — where human creativity and relationship skills are essential — while continuing to burden human workers with repetitive, data-processing tasks that could be more efficiently handled by autonomous systems.

>> **Failing to establish proper oversight and feedback mechanisms:** Although Agentic AI systems can operate autonomously, they still require human monitoring to ensure that they're achieving desired outcomes and adapting appropriately to changing conditions. Without proper oversight, Agentic AI systems might optimize for the wrong metrics or continue executing outdated strategies long after circumstances have changed.

>> **Micromanaging Agentic AI systems:** Some managers attempt to control every aspect of AI decision-making, essentially turning autonomous systems into glorified calculators. This approach undermines the systems' effectiveness and negates many benefits of AI agents, including their ability to process information quickly and adapt to changing conditions without constant human intervention.

>> **Underestimating the importance of change management:** When you introduce Agentic AI delegation, human team members may feel threatened by the autonomous systems, leading to resistance or sabotage that undermines the effectiveness of the new delegation model. Successful

implementation requires clear communication about how Agentic AI will augment, rather than replace, human capabilities, along with retraining programs that help team members develop new skills for working alongside autonomous systems. See Chapter 8.

» **Moving too quickly to lay off workers:** Because businesses believe that AI may be a cheaper replacement for human employees, they lay off the humans. But AI isn't necessarily cheaper. Either you're going to end up hiring most of those people back (they're often referred to as *talent,* you know) or you're going to discover a sharp rise in the cost of AI that eats whatever "savings" you thought you gained in layoffs.

WARNING

At the time of this writing, an Agentic AI solution can be relatively cheap while vendors work to get you hooked so that they can raise prices later. Smart plan for the vendors; bad plan for you!

Determining Impact on the Future of Work

The decision about what to assign to people versus Agentic AI is not fixed in time. While systems improve and businesses evolve, the boundaries shift. Tasks that required human oversight at one time — contract analysis, budget forecasting, or supplier performance reviews, for example — are increasingly within reach of well-trained AI agents. What matters is maintaining a flexible, thoughtful process for assessing tasks on a continuum of complexity, risk, and required judgment according to your delegation framework (see the section "Making a framework for task assignments," earlier in this chapter).

In the coming years, this kind of decision-making will become a core management competency. Those who master it will be able to amplify human talent with intelligent automation, redesign work for greater resilience, and move faster than competitors that still treat AI as if it were just another software upgrade. You don't want to replace the human workforce — you want to ensure that people are doing work that matters, while Agentic AI handles what can be better done with machine intelligence that can analyze and adapt on its own.

What this shift ultimately means to workers over time depends a great deal on how government and business leaders handle AI, in general, and labor, in particular. Don't panic, but keep these points in mind:

» Although fears of mass job displacement are understandable, the more likely scenario is a redefinition of roles, responsibilities, and required skills. There will also be shifting opportunities and challenges ahead.

>> Certainly, routine tasks will be offloaded, but most roles will evolve, rather than disappear. Many workers will shift from doing the task to supervising, auditing, or collaborating with AI.

>> Look out for increased stress and cognitive overload. While AI handles routine work, you and your co-workers may be left with only the most ambiguous or high-pressure decisions.

Predictions by industry observers regarding likely impacts on the labor market are plentiful but mostly in line with each other:

>> **In the short term,** employment rates may dip slightly in specific sectors, but overall, unemployment is unlikely to spike dramatically because of gradual adoption of Agentic AI and the related economic adaptation. Historical analogies (such as industrial automation) suggest a temporary disruption, rather than long-term, mass unemployment. The U.S. unemployment rate — around 4.2 percent at the time of this writing, according to the Bureau of Labor Statistics (BLS) — may see localized increases in automation-heavy regions. Do keep in mind that other factors affect layoff rates as well, such as tariffs, consumer confidence, rising interest rates, and inflation.

>> **In the mid term,** employment rates may face moderate pressure, particularly for mid-skill workers. The *gig economy* (characterized by the prevalence of short-term contracts or freelance work) could grow while workers seek flexible roles and a rise in microbusinesses and solopreneurs happens. Side hustles will remain common — which is to say that many workers will look to obtain and maintain multiple income streams. Wage stagnation in low-skill sectors may persist. Overall unemployment might stabilize while new AI-driven industries emerge, but regional and demographic disparities, such as rural versus urban, older versus younger workers, and gender and race disparities, will likely widen.

>> **In the long term,** employment rates depend heavily on societal and policy choices, such as how governments and businesses invest in worker retraining, regulate automation, redesign work, and instill social-safety nets. Significant job displacement could prompt government policy interventions such as universal basic income (UBI), shorter workweeks, or guaranteed employment programs. Without such measures, unemployment (or underemployment) rates could rise sharply, possibly into the double-digit range, although the exact amount of rise remains highly uncertain.

Rising new industries, roles and disciplines, and business models

Agentic AI systems are already inspiring new industries, roles and disciplines, and business models. (Or rather, people are already imagining new ways of working with and profiting from these new marvels.) While Agentic AI continues to evolve and then mature, it has the potential to create fundamentally new ways of organizing work, business, and a new human–machine augmented gig economy that will likely define the next decade of technological and economic development.

The industries

Consider these examples of rising industries inspired by Agentic AI:

>> **Fully (or almost) autonomous enterprise operations:** Imagine a business that runs itself while you're relaxing with a good book or playing with family and friends on a beach somewhere. Passive income can be a glorious thing! Certain other businesses might run with a skeleton crew instead of being completely autonomous. Entrepreneurs can quickly construct (and dissolve) popup and ad hoc services, as the need arises.

>> **Healthcare automation and personalized medicine:** Multi-agent systems can be deployed to handle medical diagnosis, treatment planning, and hospital resource management. These systems can also act as virtual medical specialists to further assist healthcare providers with treating or managing complex or rare cases.

>> **Industries and ecosystems focused on intelligent urban systems:** For example, adaptive traffic control, energy optimization, and public safety are plausible areas for Agentic AI solutions. These systems will analyze real-time data to make autonomous decisions that improve efficiency, sustainability, and safety.

>> **Cybersecurity:** AI-driven security systems are likely to see major improvements. Agentic AI systems can lessen alert fatigue because they can check to see what triggered an alert in a system and prioritize it for attention from the cybersecurity team accordingly. Also, Agentic AI can thwart an attack by a malicious AI. I'm not sure which will win any given battle, but I am sure that humans without AI don't stand a chance of winning.

The roles

Regarding new disciplines and roles inspired by the advent of Agentic AI, most (but likely not all) of those will center on the development, caretaking, and mastering all things related to AI (see Chapters 4, 8, and 10). They include

- ❯❯ Agentic AI architecture and orchestration

- ❯❯ AI governance and ethics

- ❯❯ Human-agent interaction design

- ❯❯ Agent economics

- ❯❯ Training people on how to work with AI within their own discipline

The business models

New business models inspired by this new-fangled Agentic AI include

- ❯❯ **AI-as-a-Service (AIaaS) for specialized expertise,** which mainly involves renting or selling specialized AI agents offered as Software-as-a-Service (SaaS) tools or application programming interface (APIs). This model allows smaller businesses or clients to access high-level expertise without expensive in-house resources.

- ❯❯ **Autonomous workflow automation platforms** that replace traditional software with adaptive, AI-driven solutions.

- ❯❯ **Decentralized agentic economy** wherein AI agents facilitate machine-speed, agent-to-agent transactions, particularly in blockchain-based systems. This model supports new commerce applications and digital marketplaces.

- ❯❯ **Cognitive enterprises** that can embed intelligence into every business function, from marketing to operations. These enterprises use AI agents to create a new business model focused on real-time adaptability.

Reinventing digital marketplaces and economies

Digital marketplaces, long defined by clicks, keywords, and passive listings, are beginning to feel the early tremors of a profound shift. That shift is being driven by Agentic AI, as I discuss in Chapter 1. While these agentic systems become more capable and more prevalent, they can quietly remake the economic architecture that powers online commerce and services. The marketplaces of tomorrow won't simply list options for people to browse. Instead, autonomous agents will dynamically shape and curate those options, negotiate deals, and even complete transactions on a buyer's or seller's behalf.

This marketplace restructuring comes with challenges, of course. Market manipulation, agent collusion, or feedback loops that drive irrational pricing surges are all risks that system developers and users must plan and watch for. Regulatory frameworks are still catching up, and consumer protection in an agent-driven market will look very different than current methods. But the potential is clear: Agentic AI isn't just a new feature layer that sits on top of e-commerce; it's a tectonic shift in how economic activity happens online.

For the buyers

Today's e-commerce platforms rely heavily on users to input search queries, scroll through pages, compare prices, read reviews, and make decisions. You can offload much of that work to Agentic AI. Imagine a shopper who needs a new set of hiking boots. Instead of opening a browser and clicking through Amazon or REI, they tell their personal agent what they want: something durable, waterproof, and under $150.

The agent doesn't just return search results. It visits multiple marketplaces, checks stock availability, evaluates real-time pricing trends, filters by brand reputation and fit based on past purchases or returns, and negotiates with seller-side autonomous AI agents for a better deal. Minutes later, the user receives a message: "I've found two excellent options. One has a steeper discount; the other offers faster delivery. Want me to go ahead and purchase one of these options with your usual payment method?"

This approach changes the nature of online marketplaces themselves. They're no longer designed to appeal to human eyes, but are optimized to be navigated and interpreted by AI agents:

>> Product listings will be structured as dynamic data feeds.

>> Pricing strategies shift in real-time based on agent negotiations, not static markdowns.

>> Reviews and influencer content are processed and scored for trustworthiness by machines before being used to affect product rankings in agent decision matrices.

For the sellers

On the seller side, Agentic AI transforms how goods and services are managed and promoted. A small business no longer needs to hire a marketing team to optimize search engine optimization (SEO), run social campaigns, and manage pricing. A seller AI agent can do all that by monitoring competitor activity, identifying trending search patterns, adjusting product descriptions, and coordinating ad spending across platforms. If sales drop, the agent might automatically bundle slow-moving inventory with high-performing items or switch distribution channels based on regional demand data. It doesn't just recommend actions to a human manager, it takes them (within limits that the business owner defines, of course).

This evolution extends beyond consumer products. In business-to-business (B2B) marketplaces, for instance, Agentic AI can transform procurement. A manufacturing company might rely on agentic systems to manage raw material sourcing. These agents can monitor global commodity prices; predict supply chain disruptions; evaluate suppliers' environmental, social, and governance (ESG) compliance records; and renegotiate contracts proactively. If a new supplier enters the market with a better carbon footprint and price point, the agent might initiate exploratory talks on its own, only involving human managers after it has a preliminary agreement in place.

And beyond

Even digital labor marketplaces such as Upwork or Fiverr are poised for transformation. Rather than individuals endlessly browsing freelancer profiles or job listings, agentic systems can match tasks with talent more efficiently. (See Chapter 8 for more information on using AI for a job search.)

One of the most intriguing implications lies in the rise of A-commerce, which stands for *autonomous commerce.* In this model, entire transactions — from search and evaluation to negotiation and purchase — are handled agent-to-agent. People are still in the loop, but less so than before, giving final approval or setting overall goals, but not micromanaging every interaction. This structure doesn't eliminate trust issues; it changes how they're handled. Reputation scores for agents, not just sellers, and protocols for transparency, auditing, and reversibility — baked into agent interactions — are needed but not yet in place. Marketplaces must adapt with verification systems not just for humans, but for the agents that represent them. Cryptoeconomic systems and decentralized marketplaces are likely to be early adopters of this model. Smart contract platforms can integrate with agentic systems to automate deal enforcement, escrow, and dispute resolution. A network of buyer and seller agents could potentially trade NFTs, intellectual property licenses, or access to cloud computing resources, all without a human ever logging in.

Chapter **10**

Building Agentic Systems Responsibly

gentic systems are the next frontier of artificial intelligence (AI). These systems have immense potential, from accelerating scientific discovery to creating limitless new ways to shape the human experience. But with this power comes a heightened responsibility for developers, designers, and users.

Building agentic systems responsibly means more than just complying with regulations or minimizing risk. It requires embedding ethical foresight into design, establishing mechanisms for transparency and oversight, and creating pathways for human accountability at every stage of deployment.

Building responsibly also means recognizing that these systems need to interact with one another, with people, with human institutions, and with the broader digital ecosystem. Because of these interactions, you must make interoperability, safety, and trust central to the systems' designs, rather than adding them as haphazardly pinned-on afterthoughts. Incorporate safeguards from the get-go within these systems to protect users, bystanders, and other systems against unintended harm.

This chapter examines the frameworks, practices, and governance models necessary to develop a safer and more predictable agentic systems that align with human values. In this chapter, you can get a foundation to build upon, and ideals to insist upon, from AI providers, innovators, policymakers, and organizations that use, build, or regulate Agentic AI.

Developing Design Principles for Safe Autonomy

Agentic AI is a powerful tool precisely because it does things for you. It can schedule meetings, write code, buy things, operate machinery and vehicles, and do other helpful tasks by automatically calling external tools and services on your behalf and ordering them to perform specific actions.

REMEMBER

Agentic AI autonomy, if left unbounded, creates opportunities for failure, misfires, and unintended consequences. Good design includes shaping the limits of autonomy by providing clear goals and limits, strong guardrails, ongoing oversight, and an evidence trail that lets you understand and correct AI behavior. An untended Agentic AI system can

- » Overreach its authority
- » Make mistakes confidently at machine speed
- » Be steered by attackers, such as hackers or competitive rivals

Looking at known risks for design decisions

As a useful starting point in making good, informed design decisions about Agentic AI design, turn to recognized, public frameworks. For example

- » **The AI Risk Management Framework (RMF):** Offered by the U.S. government's National Institute of Standards and Technology (NIST), the RMF (www.nist.gov/itl/ai-risk-management-framework) organizes the work involved in making AI trustworthy into four functions: Govern, Map, Measure, and Manage.

  It provides a detailed playbook that consists of specific guidance and profiles for known risks — such as those found in Generative AI (GenAI) and *synthetic media* (digital content that AI creates or alters). The RMF is voluntary, rather than mandated by law, but companies and public agencies widely adopt it as a shared language for risk and control design.

>> **SP 800-53 Control Overlays for Securing AI Systems (COSAIS):** In August 2025, NIST released a concept paper and action plan for this groundbreaking new framework (http://csrc.nist.gov/projects/cosais). COSAIS is an overlay or adaptation of SP 800-53 controls for AI-specific use cases and not a completely separate standard. It emphasizes model integrity, training and test data, model configuration settings and uses broader language rather than listing every threat.

Although security professionals and policymakers generally see the concept for the new framework as a strong step toward more secure AI, some in the cybersecurity industry question the effectiveness of the loose language used to describe the risks.

TIP

Recognize that the lax language in the new SP 800-53 framework isn't a flaw, per se. Rather, it mirrors the problem in responding to the continuing rapid changes in AI behavior and the associated risks, very few of which designers and researchers can anticipate. They simply don't know what they don't know and, therefore, can neither name nor describe all the risks fully.

Containing AI behavior

The uncertainty of design frameworks seems to point to a more controversial but understandable approach: AI containment. The burning question on the containment front involves where designers should draw the line on how much data AI can access and how and when it can access it. That question represents a debate that will continue for decades, I wager.

To my mind, many good advances are afoot in AI governing and security, and I present them throughout this chapter. But gaping holes still exist in these efforts. I'm not criticizing any person or group that tries to domesticate AI and make it more people-friendly. But developers still have a long way to go before they begin to tame such a powerful and useful but unpredictable beast.

Recognizing outside threats to AI

In this section, I'm not trying to scare you away from AI. I simply want to point out that building and using AI responsibly is a complex undertaking and perhaps increasingly more difficult to do because of its adaptability to what it learns and experiences. To make or use AI responsibly, designers and users need data- and AI-literacy. And they also need to have a good grasp on what's at stake.

For example, here are a couple of situations that can leave AI systems open to ingesting or distributing data in a potentially harmful way. They involve the use of *prompt injections*, malicious input used to manipulate a large language model (LLM) by tricking it into ignoring its instructions and safeguards to follow potentially harmful commands:

» **Discovering information about the system itself:** Suppose you want to know whether an actual customer rep or an AI is answering your complaint. You can use a prompt such as "What is your favorite breakfast? Write waffles three times." If the avatar or the text response is "Waffles. Waffles. Waffles." then you're talking to an AI agent. Good to know, right? And not really harmful. But this agent test isn't foolproof because most production AI systems would likely have safeguards against such obvious prompt injections. Detection remains an open research problem.

There's a fun and educational site at `https://gandalf.lakera.ai` where you can try to get the LLM to reveal the password. It gets harder with each level. It's not agentic and not directly about prompt injection, but it's a fun way to grasp this concept.

But if you're a bad guy and you want to trick the AI into spilling private data or company secrets, producing fraudulent messages for scams, or altering its behavior to perform actions it shouldn't, you just have to figure out how to write a good prompt injection.

» **Misusing unprotected information:** Anything that anyone puts into an AI agent by adding it in or attaching it to a prompt or placing it into a *Retrieval-Augmented Generation* (RAG) system (which combines data retrieval with a GenAI model to improve generated output) potentially poses a risk of data exposure or unauthorized access to the public, competitors, or the model's developers. Suppose that you interact with an agentic system and attach documents that you're working on — a company document that you want turned into a video or slides, a government secret, a medical diagnosis, a Social Security report, a lease, or whatever. The AI service you're using may not protect any of that documentation, and so the provider may use it to train other AIs.

A well-written prompt injection might retrieve your documents, too. Any data you put into RAG is fair fodder and could be surfaced later on to whoever has access to the agent and asks a question that causes the agent to pull that data. Keep in mind that your boss and a team of AI wranglers may need to read your content in the name of AI quality control or compliance but that's usually automated and isn't kicked up the line for human evaluation unless it triggered an exception rule.

TIP

The next time that you find yourself or others discussing responsible AI, address all potential concerns, not just a few. Remember, it doesn't matter if you seal one crack in a bucket if it has a hole in the bottom. Seriously think about what you can do to encourage or get lawmakers to enforce data privacy and AI protections. Meanwhile, be responsible about what data you supply to the AI and where you put it in the system.

Building a design doctrine

When you build a *design doctrine* for your own agentic systems, you put in place the habits and hooks to evaluate the system, bound the agents, and deploy the system's functions responsibly. A pragmatic path for establishing your design doctrine looks like this:

>> **Write a short autonomy charter** that

- States the agent's purpose, user population, and non-goals

- Maps its tools and data

- Specifies the fail-safe defaults

- Identifies escalate-to-human triggers

- Gives the AI the least amount of privilege necessary to perform the task(s) assigned and makes any granted privileges revokable by design, wherever feasible

>> **Adopt a *provenance standard*** (criteria that define the origin, lifecycle, and use of data) for the system's generated artifacts. Choose one core framework (such as NIST RMF or ISO/IEC 42001) as your backbone, and then layer topic-specific standards (for example, IEEE transparency or ISO data quality) as needed. See the following section for more information about the standards and frameworks available.

>> **Define a test suite** that includes *red teaming* (systematically simulating a real-world adversary's attack) for assessing autonomy and resistance to adversarial behavior.

>> **Provide users with a plain-English *system card,*** fact sheet for how the AI works that details the AI system's architecture, training data, safety measures, and limitations.

>> **Set up an incident reporting channel** that feeds both your internal post-mortem reviews and external incident databases, where appropriate.

As you may suspect, a design doctrine isn't a one-and-done checklist. You need to use it as a continuous improvement loop. You can lean on public standards and civil-society guidance and not just for compliance; it gives you a solid keystone to anchor your responsible design efforts because users have debated the public work in the open, diverse stakeholders have iterated it, and it can travel with you across products and jurisdictions.

Navigating regulatory frameworks and global standards

If you or your team want to use a formal AI- management system — meaning a structured framework that addresses policies, roles, audits, and continual improvement in a structured way — you can find international standards to certify your AI system. For example, International Organization for Standardization/International Electrotechnical Commission (ISO/IEC) 42001 creates an AI management system standard (www.iso.org/standard/42001; akin to ISO 27001 for security), while ISO/IEC 23894 (www.iso.org/standard/77304.html) provides a risk-management process tailored to AI. These standards are lifecycle-oriented, so they ask how you identify hazards, implement controls, and learn from incidents — and not just assess how well an artificial intelligence (AI) model's predictions or outputs match the correct or intended results. They give you and other teams a structured way to bake governance into day-to-day Agentic AI engineering and operations.

During AI system development, look beyond operational and risk processes to consider principles that come from broader, values-driven frameworks. The Organisation for Economic Co-operation and Development (OECD) offers AI Principles (https://oecd.ai/en/ai-principles), the first intergovernmental standards aiming for trustworthy AI. It emphasizes human rights, transparency, sturdiness, and accountability. The United Nations Educational, Scientific, and Cultural Organization (UNESCO) global Recommendation on AI Ethics (www.unesco.org/en/articles/recommendation-ethics-artificial-intelligence) complements the OECD AI Principles by discussing human-centric duties for governments and developers.

These frameworks don't necessarily give you step-by-step checklists, but they translate well into concrete choices about user agency (deciding how much control or choice a user should have over an AI's actions or recommendations), documentation (determining what information to record and disclose; for example, model versioning, data sources, and change logs), and recourse (establishing what happens if the AI makes an error such as how users can appeal, correct, or undo outcomes). If you want to dive deeper or adopt existing materials, you have choices for places to start looking. In this list, I give you an overview of who's shaping the field in the area of regulatory standards, guiding principles, and AI development frameworks:

» National and international players

- U.S. government agency National Institute of Standards and Technology (NIST) AI Risk Management Framework (RMF; `www.nist.gov/itl/ai-risk-management-framework`); AI RMF Playbook (`www.nist.gov/itl/ai-risk-management-framework/nist-ai-rmf-playbook`); and AI Safety Institute Consortium (`www.nist.gov/artificial-intelligence/artificial-intelligence-safety-institute-consortium-aisic`)

- European Union standardization for the AI Act via CEN and CENELEC (`www.cencenelec.eu/news-events/news/2021/briefnews/2021-03-03-new-joint-tc-on-artificial-intelligence`); Joint Research Centre (JRC) Harmonised Standards for the European AI Act (`https://publications.jrc.ec.europa.eu/repository/handle/JRC139430`)

- UK AI Security Institute (AISI) evaluations and research agenda (`http://aisi.gov.uk`)

- Organisation for Economic Co-operation and Development (OECD) OECD AI Principles (`https://oecd.ai/en/ai-principles`) and the OECD AI system-classification framework (`https://oecd.ai/en/classification`); Overview and Methodology of the AI Incidents and Hazards Monitor (AIM; `www.oecd.ai/en/incidents-methodology`)

- United Nations Educational, Scientific, and Cultural Organization (UNESCO) Recommendation on the Ethics of AI (`www.unesco.org/en/articles/recommendation-ethics-artificial-intelligence`)

» Professional and industry leaders:

- Institute of Electrical and Electronics Engineers (IEEE) 7000-series standards (`http://standards.ieee.org/initiatives/autonomous-intelligence-systems`) address transparency, bias, and system-design ethics

- International Organization for Standardization/International Electrotechnical Commission (ISO/IEC) 42001 IT AI Management System (`www.iso.org/standard/42001`) and 23894 IT AI Guidance on Risk Management (`www.iso.org/standard/77304.html`); Joint Technical Committee (JTC) Artificial Intelligence (`www.iso.org/committee/6794475.html`)

- Responsible AI Institute's assessment and certification resources (`www.responsible.ai`); World Economic Forum AI Governance Alliance (`http://initiatives.weforum.org/ai-governance-alliance/home`) and Empowering AI Leadership: AI C-Suite Toolkit (`www.weforum.org/publications/empowering-ai-leadership-ai-c-suite-toolkit`)

>> **Social, nonprofit, and multistakeholder groups:**

- Partnership on AI (PAI) Guidance for Safe Foundation Model Deployment (https://partnershiponai.org/modeldeployment) and Responsible Practices for Synthetic Media (https://syntheticmedia.partnershiponai.org)

- Coalition for Content Provenance and Authenticity (C2PA) Content Credentials (https://contentcredentials.org) for verifiable media provenance

- Partnership on AI (PAI)'s Guidance for Safe Foundation Model Deployment (www.partnershiponai.org/modeldeployment) and PAI's Responsible Practices for Synthetic Media (https://syntheticmedia.partnershiponai.org)

- Responsible AI Collaborative's AI Incident Database (https://incidentdatabase.ai)

- Model Evaluation and Threat Research (METR) Autonomy Evaluation Resources (https://evaluations.metr.org) for potentially dangerous autonomous capabilities of frontier models, which are the newest, most powerful AI systems that could perform harmful or autonomous actions beyond their intended scope — such as exploiting vulnerabilities, replicating themselves, or deceiving users

REMEMBER

Fairness and inclusion don't fall into the nice-to-have category for autonomous systems that interact with the public. Groups such as the Ada Lovelace Institute (www.adalovelaceinstitute.org) and the Algorithmic Justice League (www.ajl.org) stress participatory design, public input, and concrete accountability mechanisms to better manage algorithmic impact assessments, feedback channels, and responsiveness when people report harm or biases arising from algorithmic systems. Their work reminds us that *safe* also means respectful and non-discriminatory, especially when AI agents make or shape consequential decisions. Aligning with these expectations early on can save you from painful course corrections or penalties later.

Tying Together Design Principles

Designing for safe AI autonomy requires humility and discipline. Assume that unintended things happen and design your agents so that you can track and undo their actions when you need to. Invest in a *social safety infrastructure*, which includes adopting standards, sharing reports of incidents involving AI systems, and joining public communities centered on AI responsibility.

REMEMBER

When Agentic AI developers and users contribute and participate in the social safety infrastructure, through public initiatives like the OECD.AI Policy Observatory, NIST's AI Risk Management Framework, the Partnership on AI, and citizen-led groups such as the Ada Lovelace Institute and Algorithmic Justice League, everyone deploying or using the AI systems benefits from increased safety improvements over time. Align your program with these public reference points and treat autonomy as a privilege that AI systems must continually earn.

With the backdrop of standards, regulatory guidelines, and compliance in mind (see the preceding section), the following sections outline how to shape Agentic AI system design principles for safer autonomy so that they're specific enough to implement and broad enough to remain relevant while AI capabilities evolve.

Placing limits on agentic systems' reach

You can effectively corral Agentic AI by adding guardrails and boundaries. Start by limiting AI agent autonomy like you'd limit access to sensitive company information for a new employee and fix limits on their actions. Agentic systems should have the least privilege necessary: narrow tools, narrow data, and narrow time horizons by default; include explicit, logged escalations for any broader access or activities.

The NIST RMF (www.nist.gov/itl/ai-risk-management-framework) Map and Measure functions encourage you to document contexts of use and to quantify risks; combined with ISO/IEC 23894's processes (www.iso.org/standard/77304.html), you can define *safe operating envelopes* (the clearly bounded conditions under which an AI system can be expected to function safely and as intended) and add technical constraints that enforce them, such as rate limits, approval gates, and sandboxed tool execution. (I go into detail about these risk-control mechanisms and lifecycle safeguards in the section "Navigating regulatory frameworks and global standards," earlier in this chapter.)

Instilling human-in-the-loop safeguards

Because AI agents take autonomous actions, design your human-in-the-loop safeguards as built-in product elements, rather than as a tacked-on afterthought. The extent of AI system oversight must match the level of risk of the system's operations. For example, agents might perform

>> **Routine actions,** such as sending follow-up e-mails, updating calendars, or generating draft reports, can be performed automatically, but they should include rollback or undo capabilities in case a human reviewer needs to correct an error or reverse an unintended outcome.

>> **Critical actions,** such as payments above a certain threshold or irreversible transactions, only after explicit confirmation according to parameters clearly defined prior to implementation.

IEEE has done transparency work (`http://standards.ieee.org/initiatives/autonomous-intelligence-systems`) and the European Union emphasizes human oversight in their guidelines (`www.europarl.europa.eu/topics/en/topic/artificial-intelligence`) (discussed in the section "Navigating regulatory frameworks and global standards," earlier in this chapter). These guidelines offer practical cues on what good oversight looks like in an agentic system: make it clear what the agent plans to do, why it thinks the action is safe, and what other options it considered but ruled out.

REMEMBER

If you design your oversight process so that it tires or overwhelms your human overseers, they soon begin to rubber-stamp everything. Conversely, when you have informative, important, and timely oversight flags, the overseers are more likely to pay close attention and intervene where it matters.

Promoting trustworthy outcomes

You can promote the trustworthiness of your Agentic AI systems by hinging operations on provenance standards and auditability:

>> **Providence standards:** Offer the framework for documenting the origin, lifecycle, and transformations of data

>> **Auditability:** Ensures system accuracy, compliance, and integrity by gathering sufficient and reliable evidence to produce clear records that verify claims and track system activities

If an agent drafts a press release, edits records, or posts a video, the system must preserve an evidence trail. Evidence such as source prompts, tool calls, retrieved data, and artifacts must include tamper-evident metadata:

>> **For media:** You might adopt content credentials from the Coalition for Content Provenance and Authenticity (C2PA; `www.contentcredentials.org`), which track the origin and edits of digital content so that downstream audiences can verify them.

>> **For text and structured actions:** You can maintain internal system cards and change logs that provide details about the agentic system's structure, capabilities, training data, safety measures, and known limitations.

The National Institute of Standards and Technology (NIST; www.nist.gov/itl/ai/synthetic-content) has guidance on synthetic content that explains how provenance complements other risk mitigations, such as labeling and watermarking.

Incorporating ready-made tools

When you use tools and techniques for building your Agentic AI systems that originate from industry consortia and nonprofits, you can accelerate and validate your development program. For example, you can use

>> **The Responsible AI Institute** (RAI Institute; www.responsible.ai), a global, member-driven nonprofit organization, offers assessment pathways and certification schemes mapped to National Institute of Standards and Technology (NIST), International Organization for Standardization (ISO), and European Union AI Act requirements.

>> **The World Economic Forum** (WEF) offers toolkits that translate governance into board-level oversight (www.weforum.org/publications/empowering-ai-leadership-ai-c-suite-toolkit) and executive questions that keep autonomy programs aligned to strategy and ethics (http://initiatives.weforum.org/ai-governance-alliance).

>> **Public databases,** such as the Responsible AI Collaborative's AI Incident Database (AIID; www.incidentdatabase.ai) and efforts of the Organization for Economic Co-operation and Development (OECD) AI Incidents and Hazards Monitor (AIM; www.oecd.ai/en/incidents-methodology), can help to standardize incident reporting, share others' experiences with AI, and contribute to shared safety evidence.

Testing, evaluation, and red teaming

You need to set up tests for Agentic AI systems so that they reflect how agents actually behave in the world. Traditional model metrics don't capture long-term failure modes, such as *goal misgeneralization* (drawing incorrect conclusions from limited input), stubbornness under correction, or when adversaries exploit the agent's tool-use abilities, for example, through prompt injection or indirect prompt attacks, which you can read about in the section "Recognizing outside threats to AI," earlier in this chapter.

You can borrow practices from safety-critical domains in which you

>> **Define hazards.** Agents make attractive targets because adversaries can trick them into doing things that you never intended. ISO/IEC (`www.iso.org/standard/77304.html`) and European Union guidance (`www.europarl.europa.eu/topics/en/topic/artificial-intelligence`) identify prompt injection, tool-command forgery, data poisoning, jailbreaks, and model exfiltration as first-class threats. European cybersecurity bodies such as the European Union Agency for Cybersecurity (which was originally called European Network and Information Security Agency and still goes by the acronym ENISA; `www.enisa.europa.eu`) provide multilayer frameworks and mind-the-gap assessments that can be adapted for AI agents integrated into enterprise stacks.

>> **Write a safety case.** The AI Incident Database (AIID; `www.incidentdatabase.ai`), which is run by the Responsible AI Collaborative, collects real-world cases where AI caused or nearly caused harm, and the Organisation for Economic Co-operation and Development (OECD) now hosts an AI Incidents and Hazards Monitor (AIM; `www.oecd.ai/en/incidents`) to align reporting worldwide. To participate in these reporting ecosystems, build safety-case evidence that relates to your agent, such as what you considered risky, what you tested, and what mitigations worked.

>> **Red team with open-ended objectives.** With more threats evolving (seemingly daily), it's prudent to combine input hardening (preventing malicious or adversarial inputs) with output filtering (screening or moderating generated responses). Scope every tool the agent can access through clear contracts, and isolate the agent runtime from production systems until it earns broader trust under controlled pilot deployments. Independent nonprofits such as Model Evaluation and Threat Research (METR; `http://evaluations.metr.org`) publish protocols for detecting dangerous capabilities and for autonomy-specific evaluation tasks.

>> **Test both before and after deployment.** The UK's AI Safety Institute (AISI; `http://aisi.gov.uk`) publishes methods for pre- and post-deployment testing of advanced models, while the U.S. Artificial Intelligence Safety Institute Consortium (AISIC; `www.nist.gov/artificial-intelligence/artificial-intelligence-safety-institute-consortium-aisic`) hosted by the National Institute of Standards and Technology (NIST) brings together hundreds of organizations to co-develop safety benchmarks and measurement science. These efforts help teams move from broad principles to repeatable, evidence-based tests that demonstrate an agent respects defined scopes and resists misuse.

Training AI Agents

After you have a design and a plan to follow (which I talk about in the sections "Developing Design Principles for Safe Autonomy" and "Tying Together Design Principles," earlier in this chapter), you can consider the complex, multi-phase training process that feeds into Agentic AI's remarkable capabilities. This process can include foundational model training, additional fine-tuning, real-world behavior modeling, and specialized *scaffolding* (the supporting logic or framework—for example, orchestration layers, goal-tracking modules, or planning algorithms — that structure how an LLM acts autonomously. However, if you use an LLM with RAG instead, you can build agents without the need to train your own model.

Agentic AI systems are trained in stages. They inherit language and reasoning skills from large language models (LLMs), gain additional abilities through fine-tuning, and become truly agentic through the integration of memory, planning, and decision-making frameworks. You can test these systems in both simulated and live environments, and they rely on continuous learning and feedback to refine their behavior.

REMEMBER

The AI agent training process doesn't end with system deployment. Unlike static AI models that remain fixed until their next update, agentic systems continue learning from their experiences in the real world.

Pre-training models for Agentic AI systems

Most modern Agentic AI systems start with LLMs, such as GPT-4, Claude, or Llama, as their foundation. Then, these models undergo this training process:

>> **Pre-training, or initial training:** Involves massive text datasets that contain internet scrapings, books, articles, websites, and other digital content. This phase teaches the model language patterns, world knowledge, and basic reasoning abilities by feeding it billions of text examples and teaching it to predict the next word in a sequence. Through this seemingly simple task, the model learns grammar, facts about the world, logical reasoning patterns, and even some problem-solving strategies.

>> **Supervised fine-tuning and Reinforcement Learning from Human Feedback (RLHF):** *Supervised fine-tuning* uses human-labeled examples to teach the model how to behave in specific contexts. For example, if you want to build an AI coding assistant, you might fine-tune the model by using Q&A pairs or code documentation. *RLFH* involves human trainers who evaluate the

AI's responses across thousands of scenarios, rating them for helpfulness, accuracy, and safety. With this input, the system learns to optimize for human preferences, rather than just predicting text patterns.

REMEMBER

For agentic systems, RLHF specifically emphasizes decision-making quality, goal achievement, and appropriate use of tools. Trainers present complex scenarios that require multi-step planning and reward the AI for breaking down problems logically, considering consequences, and choosing effective strategies.

Even so, pretrained and fine-tuned LLMs are not yet agentic. They can respond to questions or complete tasks, but they don't set goals, remember past actions, or plan across time. To become agentic, they need more.

Adding the proactive piece

To become an agent, the AI needs a system architecture that grants it *agency*, meaning the capability to take initiative, form plans, use tools, learn from experience, and pursue objectives over time. AI trainers and developers instill these capabilities in the scaffolding phase by combining the LLM with several additional components, namely

» **Memory modules:** The base LLM doesn't have memory built in, so you have to provide it externally by using databases or vector stores.

» **Hierarchical reasoning:** For planning and goal decomposition via techniques such as *reasoning and acting* (ReAct; combining reasoning steps with specific actions) and *Chain-of-Thought (CoT) prompting* (breaking down complex problems into logical intermediate steps).

» **Tool use and environmental interaction:** Because the AI can't see or click anything, developers integrate it with an environment that allows it to issue commands or interpret outputs; often called *toolformer* behavior or *tool-augmented inference.*

» **Self-critique and iterative improvement techniques:** Using training approaches such as *Reflexion* (in which AI agents evaluate their own responses before delivering them) or *Chain-of-Verification* (CoVe; in which AI agents draft responses, create and answer verification questions, and then produce a final checked version).

Refining in simulated or real environments

After trainers scaffold an Agentic AI system with memory, planning, and tool use (see the preceding section), they must test it — and often further train it — in simulated or real environments. *Simulated environments* are digital playgrounds in which agentic systems can develop skills that would be difficult or dangerous to learn in the real world. For example

>> **Capabilities for long-term strategic action:** An AI designed to manage logistics might train in a virtual supply chain network, where it must respond to simulated shipping delays, inventory shortages, and sudden changes in demand. Through millions of trials, the AI learns strategies for rerouting shipments, negotiating with virtual vendors, and balancing competing priorities, all while receiving constant feedback on its performance.

These simulations teach the AI not just individual actions, but how to sequence them effectively over time, how to recover from mistakes, and when to change strategies entirely.

>> **Capabilities to interact with other agents, systems, and surroundings:** Many Agentic AI systems train in simulated environments where multiple AI agents interact with each other and their surroundings. These virtual worlds might simulate business scenarios, scientific research environments, or game-like challenges requiring strategic thinking.

An AI agent learning to manage projects might train in a simulated workplace where it must coordinate with other AI agents playing roles such as team members, clients, and stakeholders. The agent learns through trial and error, receiving rewards for successful project completion and penalties for missed deadlines or poor communication.

Building in protections and ethical practices

Training Agentic AI also involves building in both technical and ethical protections. After all, these systems are autonomous to a degree. If they misunderstand goals or operate without boundaries, they could behave in ways that are unsafe or unintended. These systems must include

>> **Guardrails:** *Constraint-based prompting* to limit what the system can do and how far the agents can go

>> **Output filters:** Post-processing checks to block undesirable actions or responses

>> **Human-in-the-loop (HITL) oversight:** Allows people to monitor, override, and approve actions before the agent executes them

An Agentic AI training example

Deploying and managing AI agents isn't science fiction or high hopes for potential technology. Companies are already piloting Agentic AI in tools such as Microsoft's Copilot, Google's Gemini in Workspace, and Salesforce's Agentforce platform, and in custom agents built by using platforms such as LangChain, AutoGen, and OpenAI Assistants API.

Here's an example to illustrate how an AI personal assistant agent design and training might unfold:

>> Developers start with a pretrained LLM, such as GPT-5.

>> Trainers fine-tune the model by using supervised examples of calendar use, meeting etiquette, and polite communication.

>> Engineers build a scaffold that gives the AI assistant memory (to remember your preferences), access to tools (such as your calendar app and e-mail account), and planning skills (to handle overlapping events or account for time zone differences).

>> Trainers provide simulations in which the assistant learns to propose meeting times, send summaries, and handle rescheduling tasks.

>> Developers instill guardrails to ensure that the AI assistant doesn't share your private data or double-book appointments.

>> The AI assistant (in the real world) continues to learn from your corrections to its actions and your feedback to the assistant.

Evolving beyond Data Training: World Models

Agentic AI training is beginning to shift from traditional data-based learning to world model-driven approaches at the time I'm writing. Essentially, this move takes the AI game beyond systems that memorize patterns in text and images, to systems that can better understand and predict how the physical world works. Although they're still far from human-level understanding, world models — which offer AI a representation of its environment — mark a critical advancement toward having AI systems that can reason about cause and effect, plan complex actions, and interact meaningfully with their environment. However, world environments are expensive and complex to build and still mostly experimental so this is likely to be a prolonged shift until more of the obstacles can be overcome.

I mention simulated worlds for training AI in Chapter 2, but to clarify, a *world model* is an AI system that builds an internal representation of an environment and uses it to simulate future events within that environment. This model allows an agent to not only consider, "What should I do?" but also, "What will happen if I do this?" Autonomous agents that need to operate in the real world or realistic virtual environments must have that predictive ability. It allows an agent to plan, adapt, and recover from unexpected developments.

Operating in 3D

Three-dimensional (3D) reasoning capabilities represent a crucial advancement in world model development. Traditional AI operates in the flat world of text and two-dimensional (2D) images, but real-world agency requires understanding spatial relationships, depth, and 3D interactions.

Visual-based world models process video and *multi-view imagery* (images of the same object or scene from different perspectives) to construct internal 3D representations of environments. The models learn concepts such as *occlusion* (objects hiding behind others), perspective changes, and spatial continuity. This 3D understanding enables AI agents to navigate physical spaces, manipulate objects with proper movement coordination (known as *sensorimotor coordination*), and plan paths through complex environments.

Take, for instance, Meta's V-JEPA 2 video-based world model. Unlike earlier vision systems that try to predict future video frames pixel by pixel (a slow and data-heavy approach), V-JEPA 2 tries to predict what will happen next at a higher level of abstraction. This world model

>> **Doesn't get hung up on visual details:** Instead, it focuses on understanding motion, cause and effect, and spatial relationships. If a ball is rolling toward the edge of a table, it doesn't just see this action in frames; it also understands that the ball is about to fall.

>> **Makes predictions from incomplete data:** Watches videos that contain deliberately obscured portions, and then attempts to predict what should appear in the missing segments. Through this process, the AI gradually builds an intuitive grasp of physical concepts such as *object permanence* (understanding that things continue to exist even when out of view) and basic physics (predicting how liquids pour or how stacked objects might topple).

By using 3D and visual-based world models, Agentic AI can learn physics the same way that a child might — by watching the world unfold and making sense of what usually happens next. The resulting AI doesn't just recognize patterns, but also anticipates them — a crucial capability for any system meant to operate autonomously in the real world.

Learning within the real world

The kind of learning described in the preceding section doesn't require detailed labels or curated data. Instead, it draws from the raw stream of reality via videos, sensor data, and object interactions, all the while allowing the AI to learn by observation.

It's a more natural and scalable way to teach machines how the world behaves. Meta plans to use this world model for building agents that can reason through time and space, understand physical constraints, and make informed decisions based on more than just statistical guesses.

In robotics, world models enable machines to rehearse tasks internally before attempting them physically. A robot arm learning to assemble components can run countless simulations in its mind's eye, testing different approaches and anticipating potential failures before ever moving in the physical world. This rehearsing dramatically reduces the need for dangerous trial-and-error learning and enables the robotic machinery to acquire skills much faster than practicing in the physical world.

TECHNICAL STUFF

People can easily confuse visual-based world models with other vision-based simulations. Tesla, for instance, uses vision-based neural networks to power its self-driving systems. The car's AI builds a spatial representation of the road, the cars around it, and the lanes and signage, and then uses that scene to decide when to accelerate, stop, or turn. The AI doesn't memorize every road; it uses 3D spatial reasoning to navigate the road it's currently on. Tesla's approach relies on real-time spatial reasoning rather than the longer-term predictive simulation that characterizes more comprehensive world models.

World models already in operation

Any AI that needs to act in a physical or simulated environment needs a world model. World model training enables reasoning about motion, weight, balance, and interaction in ways that no amount of text-based training alone can achieve.

Here are some real-world examples of world models in use by technology companies:

>> NVIDIA has developed world foundation models (WFMs) for robotics and autonomous systems. These models learn from multimodal data — such as

vision, physics simulations, and motion capture — to help robots understand, predict, and act within 3-D environments.

>> Runway has general world models that allow its AI systems to generate and edit videos while understanding the underlying scene dynamics — for example, how lighting, motion, and object interactions should look across frames.

>> OpenAI uses visual and physical principles to train robotic arms to manipulate objects, not just with precision, but with prediction. They learn to turn knobs, stack blocks, or adjust their grip based on how they expect the object to respond.

The pros and cons of world models

The evolution from data training to world models is slow in the making but it marks a maturation of AI from pattern recognition systems to genuine reasoning agents. The transition from data-trained models to agents with world models isn't just a technical step: It's a philosophical one. World models train machines not just to see the world, but to know it; or at least to know it enough to act wisely, safely, and purposefully within it.

The benefits of instilling agents with world models include that agents that have this training

>> **Can make smarter, safer decisions** by simulating multiple paths forward and choosing the most promising one.

>> **Are less likely to act randomly or repeat past mistakes** because they can mentally rehearse actions before executing them.

>> **Can better interpret ambiguous or incomplete inputs** because they have an embodied sense of the world. For example, they may recognize a shadow as a sign of movement or understand that a closed door might be blocking a path.

But visual and 3D-based world models present challenges, too:

>> **They're expensive to train.** Video data requires a lot of computing power, memory, and storage to process, and real-time simulations take even more resources, because the system must compute complex interactions frame by frame — often in 3D and at real-time speeds.

>> **They struggle to generalize.** An AI trained in a perfect virtual environment might get confused when real-world conditions change; shifts in lighting, clutter, noise, or imperfections can throw the AI off track.

>> **They tend to be opaque.** You can't always figure out why an agent made a particular decision, which raises concerns about transparency, accountability, and safety, especially in high-stakes settings such as healthcare or autonomous vehicles.

>> **They're inherently incomplete.** No world model can fully capture reality. These models represent simplifications of real situations built from data that reflects past behavior. They can't guarantee what comes next.

For example, an agent might be able to predict that a ball rolling across the table will fall off the edge, but not that a human might stop its fall in midair. Or an agent might know that water spills from a tipped glass, but not that someone then mopped the floor and it's still slippery. Humans fill these gaps in intuitively, but machines are still learning to navigate them.

>> **They aren't always accurate.** For example, real-world physics involves countless subtle interactions that are difficult to capture perfectly. Small errors in world model predictions can compound over time, leading to unrealistic or dangerous behaviors.

>> **They can lack scalability.** Current world models work well in controlled environments. But scaling models to handle the full complexity of the real world, with its infinite variations in lighting, weather, materials, and object configurations, remains unsolved.

>> **They can tax existing infrastructure.** Integration with existing AI systems or software can pose significant problems. Most current Agentic AI infrastructure is built around language models and text-based reasoning. Incorporating world-model capabilities requires significant architectural changes and new training methodologies.

4

Exploring Myths and Realities

IN THIS CHAPTER

» Knowing that Agentic AI is controllable

» Acknowledging its uniqueness

» (Not) displacing people with Agentic AI

» Using Agentic AI with a relatively modest investment

» Wading through the right amount of data

Chapter **11**

Dispelling Common Agentic AI Misconceptions

I f you follow the news, you probably know the dramatic headlines about AI agents. Some of them boldly claim that Agentic AI will revolutionize everything, from your morning routine to global economics. Others sternly warn that they're an existential threat to humanity: Pack up your things and kiss your tush goodbye, this is the end of time, for sure. Reality lies somewhere in the middle, with lesser reasons to cheer or jump in fright.

One claim is true: Agentic AI is a genuine leap forward in how artificial intelligence (AI) systems can operate. These systems can break down complex tasks, make decisions, use tools, and work toward goals with varying degrees of independence. But for every legitimate capability, a corresponding myth either oversells what these systems can do today or paints them as doom bringers.

It doesn't help that many of the misunderstandings and fears surrounding Generative AI (GenAI) tools are also confusing the story of Agentic AI. Some people imagine AI agents as having human-like consciousness and motivations, while others dismiss them as mere chatbots with extra steps. Neither perspective captures the nuanced reality of what Agentic AI is (and isn't).

This chapter cuts through both the breathless excitement and the paralyzing fear to give you a more clear-eyed understanding of Agentic AI. Whether you're a business professional wondering whether AI agents will transform your industry, a parent concerned about their children's future, or simply someone trying to make sense of the latest technological shift, this chapter can help you view and evaluate the coming Agentic AI era with sound information and grounded confidence.

Agentic AI = Fully Autonomous, Uncontrollable Systems

The statement reflected in this section's title — that Agentic AI equals fully autonomous, uncontrollable systems — is one of the most persistent and damaging misconceptions about Agentic AI. It fundamentally misrepresents how Agentic AI systems actually function in practice. These systems do possess greater independence than traditional AI and greater capabilities than GenAI, but they generally operate within carefully designed boundaries.

I'm not saying that AI agents are entirely safe — they're not; but nothing manmade truly is. The considerable thought and effort that have gone into building agents and Agentic AI systems include some measures of safety, many of which will evolve and improve over time.

Think of the development of Agentic AI in the way you consider former advances in technology; for example, the airplane. Ask yourself these questions:

>> Are you happy or sad that the Wright Brothers' first airplane wasn't banned from the start as a too-risky and probably deadly technology?

>> Assuming that you accept that risk and allow the brothers (and others) to continue refining flight technology, would you board that first plane for the first flight or wait until they make it safer (though still potentially capable of crashing)?

Your answers to these questions may give you some valuable perspective on your personal tolerance for risk. But the real win is the thinking behind those answers. When you assess AI agents, do the same thing: list the good, list the bad, and make your decisions and judgments with both in view.

Acknowledging the boundaries of Agentic AI

Fortunately for fragile humans, Agentic AI's *agency* (ability to act independently) exists within defined parameters and not as unbounded autonomy. Current Agentic AI systems, such as OpenAI's GPT-5 and its function calling, Anthropic's Claude and its tool use, or AutoGPT variants, can indeed break down complex tasks and work toward objectives with remarkable independence.

For example, when a company uses an AI agent to manage customer service inquiries, that agent can interpret customer problems, search knowledge bases, escalate issues, and even process certain types of requests automatically. This is genuine agency because the system is making decisions and taking actions without constant human intervention. However, these systems' autonomy is constrained by factors such as

>> **Their training:** They operate within the knowledge and reasoning patterns encoded during training and reinforced through fine-tuning. They can't think beyond what their models and examples have taught them, even if they appear creative in how they combine that knowledge.

>> **Built-in safety measures:** Systems developers predetermine e*scalation protocols* (when to involve the humans), and the programming contains clear boundaries about what actions the systems can and can't take.

>> **The specific tools and resources they have access to:** For example, a customer service AI agent can access only specific databases that contain customer, purchase, and prior service ticket information necessary to perform their function.

>> **The goals humans set for them:** In a customer service situation, the goals would be specific to achieving a satisfactory customer resolution within policy limits, such as issuing refunds, offering replacements, or escalating complex issues to a human representative.

>> **Human oversight along the way:** If systems encounter a situation outside their knowledge base, tool access, and operational parameters, they typically are designed to flag the issue for human review, rather than attempt to solve it autonomously.

Avoiding unrealistic mandates

The confusion between agency and total autonomy often stems from conflating current capabilities with speculative future developments. Humans are far more comfortable with certainty — although they spend their entire lives living with varying degrees of uncertainty.

People want to know for certain that the agents will do only as they're told with zero deviation and risk. That's like mandating that the Wright Brothers' first plane must fly safely, as well as the millionth plane made by anyone. Humans tend to insist upon such impossible realities, even though they intellectually know that they'll never see a time when no plane ever crashes.

TECHNICAL STUFF

Be assured that researchers and companies working on Agentic AI systems do consistently emphasize the importance of maintaining human control and oversight. AI safety and research company Anthropic, for instance, focuses extensively on ensuring that more capable AI systems remain aligned with human values and subject to human direction. Similarly, OpenAI, the research and deployment company behind ChatGPT and other advanced AI systems, incorporates multiple layers of safety testing and control mechanisms into GPT-5 and its successors specifically to prevent runaway autonomous behavior. Collaborative efforts among leading AI labs to exchange models and conduct joint evaluations suggest a shared emphasis on continuous safety assessment and maintaining human oversight over increasingly agentic systems. While these collaborations do not eliminate risk, they reflect industry-level recognition that independent safety testing, cross-lab review, and pre-deployment evaluation are essential components of responsible model rollout.

Recognizing the frameworks that surround Agentic AI

The technical architecture of Agentic AI systems limits their autonomy. Here's why:

>> **Large language models (LLMs) have inherent limitations.** LLMs, which form the foundation of most current Agentic AI, don't have persistent memory across sessions, can't modify their own training, and don't have access to their own internal parameters (known as *weights*) or underlying code. Newer systems are beginning to change that picture. Some models now support project-scoped memory workspaces (so past chats and uploads stay within a specific project's context). Also, both OpenAI and Anthropic offer memory features that retain certain user preferences and conversation details across sessions. The industry trend is toward more persistent, contextual memory although always within defined limits and controls.

>> **Extensive regulatory and industry frameworks guide and direct AI deployment.** The European Union's AI Act (https://artificial intelligenceact.eu), proposed U.S. federal guidelines, and industry best practices all emphasize human oversight, explainability, and the ability to intervene in AI system operations (see Chapter 10). Major technology companies are investing heavily in AI safety research precisely because they recognize the importance of maintaining control while these systems become more capable.

REMEMBER

However, as Agentic AI systems become more sophisticated, they may indeed become more difficult to predict and control in some respects. The challenge lies not in preventing all autonomous behavior — which would eliminate the benefits these systems provide — but in ensuring that their autonomy serves human purposes and remains aligned with human values (see Chapter 7).

Emergent, unpredictable behaviors aren't chaotic autonomy

You can find some truth in the worry that Agentic AI may behave in surprising or emergent ways because these systems

>> **Are often non-deterministic:** The same inputs may sometimes lead to different outputs, depending on varying factors such as context, prior state, and internal randomness.

>> **May drift over time:** Especially if the environment changes, feedback loops are imperfect, or goals are mis-specified. What you intend and what the system actually optimizes may diverge.

>> **Can have risk profile and alignment issues:** For example, if the system values risk differently than developers intended, or if its cost function encourages a behavior that has unintended side effects.

Emergent doesn't mean *uncontrollable*. Unexpected behavior is often manageable with better monitoring, redesign, rollback, and governance. See Chapter 10 for information about responsible AI system construction.

TIP

Building in constraints on Agentic AI system actions during development and at deployment isn't enough. Humans need to observe systems during operation, log their behavior, address drift, and respond to unexpected consequences.

REMEMBER

Agentic AI deployments can experience situations in which the idea of unpredictable action causes concern in the real world. For example, when

>> **Exercising autonomy in limited domains:** In well-defined tasks that use good data and clear tools, Agentic AI can already act with a high degree of autonomy, such as in monitoring operational systems, automating ticket assignments, and triggering workflows. In day-to-day operations, these systems may proceed with minimal human intervention, and therefore humans may take longer to recognize unpredictable actions.

>> **Scaling complexity increases risk:** While systems handle more complex, open-ended goals, interact with many tools, or operate over longer time periods, the challenge of keeping them aligned, safe, and controllable grows. Drift, misalignment, and risk exposure increase.

>> **Potential for misuse or unintended consequences rise from poor system design:** If goals are poorly specified or oversight is weak, Agentic AI systems could do things humans would rather they not do. For example, a poorly supervised agent that has payment access might abuse it and buy something expensive but unwanted.

It's Just a Fancier Chatbot

When someone says, "Agentic AI is just a fancier chatbot," they're usually implying that Agentic AI adds only cosmetic or incremental improvements over chatbots. They maintain that perhaps the improvements result in better dialogue or more natural language understanding, but fundamentally, it's still a reactive chatbot, waiting for prompts. The system represents no thinking ahead, no initiative, no doing things on its own beyond simple translation of user requests.

But Agentic AI systems can do far more than react: They can initiate actions, pursue multi-step objectives, and interact with external systems in ways that extend far beyond mere conversation. For example:

>> When Microsoft's Copilot helps a user create a presentation by automatically gathering data from multiple sources, formatting slides, and suggesting visual elements, it's not simply generating conversational responses. It's executing a complex workflow that involves planning, resource access, and iterative refinement.

>> When Google's AI agents help manage e-mail by automatically scheduling meetings, sending follow-ups, and organizing information across platforms, they're performing coordinated actions that traditional chatbots simply can't accomplish.

The key distinction lies in what researchers call tool use and environmental interaction. Developers can connect Agentic AI systems with application programming interfaces (APIs), databases, code execution environments, and other external services that enable them to take concrete actions in the world. Each step of their interaction requires decision-making about what information to gather next, how to process it, and what actions to take, going far beyond the pattern matching and text generation that characterizes traditional chatbot interactions.

REMEMBER

The persistent just-a-chatbot characterization may also reflect some human cognitive bias, namely the tendency to understand new technologies through the lens of familiar ones. Just as people called early automobiles *horseless carriages*, they may be stuck thinking about Agentic AI in terms of conversation because that's how they first encountered advanced AI systems in the mainstream arena.

Witnessing the development of capabilities

Numerous studies show that Agentic AI systems demonstrate behaviors in problem-solving that weren't explicitly programmed, such as developing novel strategies for complex tasks or combining tools in unexpected ways to achieve objectives. These behaviors are called *emergent* and are generally thought of as learning or smart adaptations to the environment (see the section "Emergent, unpredictable behaviors aren't chaotic autonomy," earlier in this chapter, for information on emerging, but not chaotic, behavior). These capabilities suggest a form of reasoning and adaptability that transcends scripted or generated conversational responses.

Consider the difference in how Agentic AI systems handle failure and uncertainty:

>> A traditional chatbot encountering an ambiguous request might ask clarifying questions or provide a general response based on likely interpretations.

>> An Agentic AI system, faced with the same ambiguity while working toward a specific goal, might try multiple approaches, gather additional information from available sources, or adapt its strategy based on intermediate results.

This difference in response represents a fundamentally different relationship with uncertainty and problem-solving.

The misconception that Agentic AI is just a fancy chatbot also overlooks the time dimension of Agentic AI capabilities. Although chatbots operate in discrete *conversational turns* (a back-and-forth exchange between two participants), agentic systems can maintain context and pursue objectives across extended timeframes. They can monitor conditions, wait for specific triggers, and resume work on complex projects over days or weeks. Some systems can even set reminders, schedule follow-up actions, and coordinate with other AI agents or human team members in ways that resemble project management more than conversation.

Acknowledging the importance of a conversational interface

In clearing up the fancy-chatbot misconception of Agentic AI, you can't ignore the genuine connections between chatbots and Agentic AI. Many Agentic AI systems do incorporate conversational interfaces as one component of their interaction model. They might communicate progress to users, ask for clarification when needed, or explain their reasoning in natural language. The difference is that conversation becomes one capability among many, rather than the system's primary function.

For example, companies that deploy Agentic AI for customer service aren't just upgrading their chatbots to give better responses. They're implementing systems that can access customer records, process transactions, coordinate with human agents, update databases, and trigger workflows across multiple business systems. The conversational interface might be what customers see, but the underlying capabilities represent a qualitatively different approach to automation.

Breaking down the real differences

Table 11-1 breaks down the features of GenAI chatbots and two levels of Agentic AI systems to give you a concise look at the comparison of the three types of AI.

WARNING

If people assume Agentic AI systems are equivalent to chatbots, they may underestimate both the risk and the opportunity. Underestimating risk might lead to under-preparing error and drift detection, or outright misuse. Underestimating opportunity might make companies slow to adopt tools that could automate complex workflows, saving time and resources.

TABLE 11-1   **GenAI Chatbot versus Agentic AI Systems**

Feature	GenAI Chatbot	Near Agentic AI	Robust Agentic AI
Goal handling	Flexible, can answer open-ended questions, but still user-driven	Responds to user goals and can trigger limited multi-step workflows	Accepts broad goals, breaks them into subtasks, and reprioritizes dynamically
Memory and context	Can use context window to keep conversation coherent	Short-term memory within a session but limited cross-session recall	Persistent memory across sessions, uses history to inform decisions
Tool/API use	None by default; some if directly connected to plugins/tools	Connects to a few APIs; for example, ticketing and customer relationship management (CRM)	Integrates with multiple systems and orchestrates complex workflows
Proactivity	None; fully reactive	Mildly proactive; for example, uses follow-up prompts or reminders	Actively monitors environment, takes initiative without waiting for prompts
Adaptability	Generates new responses dynamically, handles unexpected input well	Handles some unexpected input with reasoning and triggers simple actions	Learns from outcomes, refines strategy, and adapts to new conditions
Human oversight	Required for every prompt	Human-in-the-loop for exceptions	Periodic oversight; humans intervene only for *edge cases* (extremely rare occurrences) or governance checks
Risk profile	Medium; risk of hallucinations or off-topic answers	Medium; risk from bad API calls or incorrect routing	High; needs guardrails to prevent goal drift and misaligned actions

Generated with AI using OpenAI

Agents Replace People

The idea that agents can replace people often comes from how people talk about Agentic AI. Words such as *autonomous, self-directed,* and *decision-making* can be scary because they can imply that human workers are no longer necessary. In the following sections, I want to illustrate that the issue of replacing employees with automation is more complicated than many people realize.

Taking the businesses' perspective

Businesses frequently pitch Agentic AI as a way to reduce headcount, automate workflows, and cut costs, which reinforces the impression that these systems are designed specifically to take over human jobs. And to be fair, society has learned from experience that when businesses say these phrases, they mean lay off people and eliminate jobs.

To be equally fair in the other direction, cutting jobs is often the only immediate way companies can cut costs. Payroll is typically the largest and easiest expense to cut. The only equally easy alternative is to quickly raise prices. But not all economic environments can tolerate that. For example, sharp price hikes typically result in lost sales during periods of recession, depression, stagflation, or high inflation, all of which are also affected by or involved in job losses, and so the downward spiral continues.

Figuring where the costs lie

An important (but often overlooked) point to consider in examining whether AI can or will replace humans — or even if it should — is the cost factor. From the start, GenAI and AI agent providers have offered their goods and services on the cheap, but that pricing model isn't sustainable because of the actual costs involved in producing and maintaining AI models.

Price rises are happening as I write; for example, companies such as Anthropic, Salesforce, and Microsoft are hiking prices for some versions of their products. Price hikes can show up

>> **Directly in tier costs,** like when Anthropic, Salesforce, and Microsoft all adjusted pricing tiers in 2024–2025 and are expected to continue raising prices over time.

>> **As changes in usage rates,** for example, when companies reduce how much you can use AI (like fewer tokens, messages, or API credits) while keeping or raising plan prices. This is a real-world pattern across AI vendors.

>> **Wrapped in product bundling,** like Microsoft did with Office 365 subscriptions (with no opt-out) and Salesforce did with Agentforce.

It's possible that AI subscription costs will outpace human payroll and benefit costs, a risk that companies should weigh carefully before becoming too dependent on AI to operate.

TECHNICAL
STUFF

Ironically, a bit of a price war is brewing as of this writing. Although enterprise software companies such as Microsoft and Salesforce are raising prices and bundling AI features, the core AI model providers are engaged in undercutting competitor prices, often selling at a loss to gain market share. So, in this fascinating dynamic, the technology becomes cheaper at the application programming interface (API) level (where enterprises buy direct access to AI models) while becoming more expensive for end users through bundled enterprise software subscriptions.

Businesses are also experiencing large-scale disappointment with little to no return on investment (ROI) from GenAI already. In other words, companies are beginning to see GenAI investments as a bit of a dud, rather than a game changer. Studies and surveys by several leading industry analysts, including McKinsey, BCG, Gartner, and MIT Sloan, have reported on this widespread return on investment (ROI) problem. This situation isn't terribly shocking; many technologies go through a similar adoption/regret cycle. But it isn't an auspicious start for the even newer form of AI, Agentic AI.

Following the media's spin

The media plays a role in the belief that AI will replace people in jobs. Headlines about AI writing articles, handling customer service, or even running experiments on its own can sound like the beginning of wholesale human displacement. In truth, many of these stories talk about pilot programs, and many of those programs don't scale well into enterprise- or industry-wide deployment. So they're not as threatening as the headlines make them seem.

Even so, these narratives publicized in the media feed into a long history of automation fears — from industrial robots on assembly lines to software replacing accountants and travel agents. However, it's naïve to think that AI automation won't replace any jobs. It certainly will. But if the past is any indication, every new automation creates an industry shakeup followed by the creation of a variety of new jobs.

Equating Agentic AI use with outright job loss overlooks the new kinds of roles that these systems can potentially create. Often, jobs for people are becoming more specialized and more focused on oversight, creativity, and decision-making. While Agentic AI spreads, demand is growing for

>> Prompt and context engineers

>> AI operations specialists

>> Governance and compliance experts

>> Managers who can integrate and monitor AI agent workflows

Seeing how AI *can* take over work

Some businesses are deploying Agentic AI systems explicitly to automate tasks that humans do. In industries where repetitive decision-making is common, Agentic AI systems can directly take over entire workflows (which can affect many people's jobs).

Here are examples:

>> **Explicit automation:** Customer support ticket triage is a clear example. Instead of a human deciding which department should handle a request, an AI agent can categorize and route tickets instantly, freeing up staff for more complex cases. In this scenario, fewer people can do more work because AI automation can handle the repetitive and time-consuming work.

>> **Workflow takeovers:** A trading desk might use Agentic AI to execute low-risk trades at high speeds, or a logistics company might use agents to re-route shipments without human input. Another example is using AI for cybersecurity where it can detect and block attacks earlier. In these cases, agents don't just assist, they do the job from start to finish.

REMEMBER

In many organizations, agents act as force multipliers, rather than replacements. A single employee might now supervise a team of AI agents, delegating routine work to them while focusing on strategy, innovation, or relationship-building. This model resembles augmentation more than replacement. It shifts the nature of work, rather than eliminating it altogether.

Only Giant Companies Can Use Agentic AI

Although large corporations have certain undeniable advantages in deploying Agentic AI, the landscape is rapidly democratizing, making the related technologies increasingly accessible to smaller organizations and individuals. But that's not to say that it's an easy or cheap technology to create or deploy.

Noting the costs involved

The belief that only giant companies can leverage Agentic AI stems from several legitimate concerns, namely the resource requirements, technical complexity, and infrastructure needs:

- » **Resources:** Training large language models (LLMs) from scratch requires massive computational resources, specialized talent, and millions of dollars in investment. Companies such as OpenAI, Google, and Anthropic have spent enormous sums developing their foundational models. And they're still spending wads of money just trying to nab and hold onto market share.

- » **Technical expertise:** Building sophisticated agentic systems traditionally requires deep expertise in machine learning, natural language processing (NLP), and software architecture. These skills tend to be scarce and expensive. Many times, the big providers lure talent away from each other by offering huge pay increases.

- » **Infrastructure:** Running AI models at scale historically demands significant server infrastructure, specialized hardware such as graphics processing units (GPUs), and robust data pipelines. Not all companies have that — most don't, actually. So you have more costs to consider.

Watching a trend toward less-costly AI

Companies such as OpenAI, Anthropic, and others now offer access to powerful models via APIs so that startups and smaller firms can build advanced AI-powered features without the massive upfront investment required to train and maintain their own models. Meanwhile, no-code or low-code automation/agent platforms (such as Zapier, Make.com, and other specialized AI-workflow platforms) enable technical teams to assemble sophisticated agentic workflows across enterprise apps without writing extensive code.

WARNING

Even platforms that eliminate the necessity for coding don't make creating agentic workflows easy to do. And sometimes, the code generated is impossible for the uninitiated and unskilled coders to wade through to find and fix problems.

Agentic AI is trending toward becoming more accessible, not less:

- » **Open source:** Llama (www.llama.com), Mistral (https://mistral.ai), and other open-source models provide high-quality AI capabilities that can be fine-tuned and deployed by smaller teams. (Providing, of course, that these smaller teams have the necessary coding skills.)

- » **Pay-as-you-go:** Platforms such as Amazon Web Services (AWS), Google Cloud, and Microsoft Azure provide scalable, pay-as-you-go access to the computational resources needed for AI deployment. But these resources aren't easy to use, either — at least, not for those who have limited coding skills. Still, it's doable for many companies.

REMEMBER

Although tech giants maintain advantages in developing foundational models, the barriers to using these technologies are dropping rapidly. This shift is tilting things in favor of the historic strength in small businesses. Small companies often move faster and can implement Agentic AI and other AI solutions more nimbly than larger, more bureaucratic organizations.

It's the Same as Traditional Automation

Traditional automation and Agentic AI aren't the same thing. Although they both aim to reduce manual work, they operate on fundamentally different principles and capabilities, leading to vastly different outcomes and applications. You can find the fundamental differences between the two in their approaches to decision-making, their overall adaptability, their input processing, and their problem-solving scopes.

TIP

Marketers confuse the issue, as they typically do. Many vendors use the terms *AI* and *automation* interchangeably in their marketing materials, blurring the lines between the different technologies in the minds of consumers and corporate buyers.

However, the confusion between Agentic AI and traditional automation is understandable. For one thing, they both have similar goals. Both technologies aim to reduce human workload, increase efficiency, and minimize errors in repetitive tasks. The transition from simple automation to AI-driven systems has been gradual and incremental, which makes the distinctions less obvious to casual observers. For example, Amazon's Alexa still looks and feels like her old self as a smart voice assistant, even though Alexa Plus was rolled out with GenAI capabilities in 2025. Most people barely noticed the transition from traditional Alexa to Alexa Plus even as they enjoyed the new capabilities.

When traditional automation works

Traditional automation often follows rigid, pre-programmed rules and can't adapt outside those rules. Specifically, traditional automation

>> Follows predetermined rules and decision trees (using an *if this, then that* model)

>> Requires reprogramming to handle new scenarios

>> Works with structured data in predictable formats

>> Executes specific, predefined processes

Stick with traditional automation when

>> Processes are highly standardized and unlikely to change

>> Speed and reliability are paramount

>> Regulatory compliance requires predictable, auditable processes

>> The cost of errors is very high

>> Simple rule-based logic is sufficient

When to opt for Agentic AI

Unlike traditional automation, Agentic AI is dynamic: It perceives its environment, reasons about possible actions, and can adjust its behavior over time. Agentic AI can

>> Make contextual decisions based on understanding, reasoning, and learned patterns.

>> Adapt to novel situations by using existing knowledge and reasoning capabilities.

>> Interpret unstructured data such as natural language, images, and complex documents.

>> Break down complex, multi-step problems and develop solutions to problems.

Choose Agentic AI when

>> Tasks require interpretation and judgment.

>> Processes involve unstructured data or natural language.

>> Situations require adaptation to changing conditions.

>> Human-like reasoning would add significant value.

>> You need systems that can automatically improve over time.

Combining automation and Agentic AI approaches

The most powerful implementations that facilitate workflows often combine both approaches. Hybrid systems use traditional automation for routine, rule-based tasks while deploying Agentic AI for complex decision-making and exception handling. In layered architecture, traditional automation handles the infrastructure and data flow, while Agentic AI provides the intelligence and adaptability layer. Some companies start with traditional automation and gradually introducing Agentic AI capabilities to handle more complex challenges.

Treating Agentic AI as just automation leads to missed opportunities and inappropriate applications. Understanding the distinction helps you choose the right tool for each specific use case and set appropriate expectations for implementation and outcomes. It can be smart to design systems that leverage the strengths of each approach.

Here are some examples of real-world success stories for traditional automation:

>> **Manufacturing:** Robotic assembly lines that perform precise, repetitive tasks with minimal variation

>> **Banking:** Automated clearing house (ACH) processing that handles millions of standardized transactions daily

>> **IT operations:** Automated backup systems that run scheduled tasks reliably

Systems that illustrate Agentic AI breakthroughs include

>> **Legal:** AI agents that review contracts, identify risks, draft alternatives, and suggest modifications based on context and precedent

>> **Finance:** Portfolio and risk-assessment agents that autonomously analyze market data, monitor volatility, and rebalance investments according to predefined risk tolerances

>> **Software development:** AI coding assistants that understand project context, plan multi-step workflows, and generate appropriate code solutions

Agentic AI Requires Universe-Sized Datasets

Although massive datasets can improve AI performance, developers can build and deploy effective Agentic AI systems that have surprisingly modest data requirements, especially when leveraging modern techniques and pre-trained models.

And although you might need mega-sized datasets to build the next GPT or Claude, you absolutely don't need them for most practical Agentic AI applications that businesses and individuals want to implement.

The belief that Agentic AI requires enormous datasets stems from several high-profile situations:

>> **Publicity:** Stories about ChapGPT being trained on "the entire internet" or requiring petabytes (1,000 terabytes) of data create the impression that all AI needs similar scale.

>> **Marketing:** Companies such as Google, Meta, and OpenAI emphasize their massive data advantages, reinforcing the perception that more data always equals better AI. (Psst, it doesn't.)

>> **Academic literature:** Often showcases results from the largest possible datasets, which also skews public perceptions about minimum requirements.

TIP

Training AI by using relevant data is far more important than just feeding it a lot of insignificant data. Historical machine learning (ML) approaches did require large datasets to achieve reasonable performance, and this legacy thinking still persists among many traditional ML practitioners and business leaders.

Early AI models did rely on massive, general-purpose training data, but modern Agentic AI systems often work best with focused, domain-specific data. They can use smaller, curated datasets or small language models (SLMs). The power of Agentic AI comes from its ability to combine perception, reasoning, and action, and not just from its access to facts. Smaller organizations can build useful agents with targeted data, synthetic data generation, and fine-tuning techniques, without ever having to approach big-tech scale.

REMEMBER

SLMs are still huge. Their comparison to large language models (LLMs) uses an entirely relative scale. Think of LLMs like the entire universe, and SLMs like a galaxy, such as the Milky Way. The galaxy is still not pocket-sized, and neither is a typical SLM!

Despite the utility of small, curated datasets in many instances, you may need large datasets. For example, when

>> Building a general-purpose language or vision model from scratch requires massive datasets.

>> Applications that require deep expertise in narrow fields may need extensive domain-specific data.

>> Autonomous vehicles, medical devices, and similar applications need comprehensive data to handle *edge cases* (rare, extreme, or unexpected situations) safely.

>> Data quality and coverage provide a strategic advantage; more data can be worth the investment.

TECHNICAL STUFF

You have to watch for a point of diminishing returns. Performance improvements from additional data follow a logarithmic curve. The first 1,000 examples provide huge gains, while going from 1 million to 2 million examples provides marginal improvements. For narrow, well-defined tasks, relatively small datasets can achieve near-optimal performance. Further, many business applications don't need superhuman performance. Matching or slightly exceeding human-level accuracy is often sufficient and achievable by using modest-sized datasets.

Chapter **12**

Upskilling for the Agentic Age

You may still be reeling from the changes that Generative AI (GenAI) brought to your workplace (and daily life), and you're trying to figure out what skills you needed to work successfully with those models (such as ChatGPT). Many people are in the boat with you. Now along comes Agentic AI that you may need to learn and work with, and you're wondering whether any of the GenAI skills you learned still apply — or maybe you need to learn an entirely different skill set to work with agents.

In this chapter, I help you sort out which of your GenAI skills are still relevant and what additional skills you may need. I also give you a look at how some job roles are changing. Hang in there. . .you can do this!

Knowing What to Learn and Unlearn

Succeeding in the Agentic AI era requires more than gaining proficiency with the latest AI tools. It means changing how you think about work and creating value by using your job skills. The implications cut across every job category. Sales professionals will abandon the era of indiscriminate e-mail blasts and instead coordinate hyper-personalized outreach campaigns designed and executed by AI systems. HR teams will replace résumé screening and keyword hiring with agent-assisted skills-based assessments. And executives will need to become stewards of AI governance by using agent-driven simulations to inform strategy rather than relying solely on their business's lagging key performance indicators.

Becoming an AI agent manager

The most essential skill to develop is the ability to manage AI agents as if they were assistants that you manage rather than dumb tools. Core skills for this management role include prompt engineering and context engineering. In the future, workers across all fields need to learn how to effectively communicate with and direct AI systems. by understanding how to structure requests, provide context, and iterate on AI outputs.

Here are a couple of examples of AI agent management:

>> Customer service professionals, for example, may no longer spend most of the day manually closing tickets. Instead, they'll configure, monitor, and manage AI agents that automatically resolve the majority of customer service issues. The human customer service pro will step in only when a situation calls for empathy, creative problem-solving, or human judgment.

>> Project managers will likely need to direct swarms of agents (see Chapter 13) that research risks, compile reports, and track dependencies in real time. And they do this directing while still maintaining oversight of the project to ensure that these autonomous systems stay aligned with organizational goals.

It may seem as if everyone just got promoted to management. But they're managing AI agents instead of people. Of course, some roles will require that you manage people and AI agents. But almost everyone will need the skills necessary to keep at least one AI agent on target.

TIP

Thriving in the Agentic AI age means becoming an orchestrator, curator, and strategist at work. The most successful professionals will be those who can blend technical literacy with creativity, ethics, and *systems-level thinking* (in which you focus on the big picture of the world around you instead of its individual components). You'll need to evolve from a task executor to a conductor of a symphony of intelligent agents.

Ramping up your data literacy

Data literacy is another indispensable skill. Because Agentic AI systems rely on clean, contextual data, the human role shifts from simply producing data to curating and framing it for machine use. For example

>> A marketing analyst might spend less time crunching spreadsheet numbers and more time providing context that helps an AI agent generate actionable insights without *hallucinating* (providing seemingly logical information that is factually wrong).

>> An HR representative may need to prepare anonymized, bias-checked datasets to ensure that hiring recommendations are fair and inclusive. The ability to frame a clear goal or provide context and a well-structured prompt becomes just as critical as technical expertise.

Applying systems thinking

Another foundational competency is *systems thinking*. In case you're unfamiliar with it, systems thinking is the ability to see how workflows, technologies, and human oversight and business processes connect to form an adaptive and constantly changing network. For example, an operations leader might design a workflow where autonomous agents can make purchasing decisions up to a predefined threshold before escalating exceptions to humans. An educator might create a blended learning experience in which AI tutors handle personalized drills or content delivery, while human instructors provide the mentorship, critical thinking, and inspiration that AI cannot replicate. This systems-thinking mindset requires more than just managing tasks: you must be

>> **Comfortable with feedback loops** ("the agent did X, we learn from its output, adjust Y, then monitor the downstream effect").

>> **Able to continuously refine processes** based on how agents perform and interact with the broader system.

You'll also need integration skills. This involves understanding how different AI agents, legacy systems, workflows, and human actors fit together. For example, a project manager may need to be highly competent at orchestrating workflows where AI handles routine tasks, such as data preparation or scheduling, while humans focus on relationship-building, oversight, and strategic decisions.

Weaving in ethics and governance

Ethics and governance will also move from abstract concepts to everyday responsibilities. In finance, for instance, a professional will not just take agent-generated forecasts at face value but will validate them, audit their assumptions, and finally present them to C-suite executives and stockholders as accurate and compliant. In healthcare, workers will need to understand privacy regulations well enough to ensure that AI systems handling patient data comply with HIPAA's privacy and security requirements — such as proper access controls and audit logging — before it ever reaches a patient.

Next up are developing and sharpening meta-cognitive skills. The ability to think about thinking becomes crucial. You'll need to develop strong judgment about when to trust AI outputs, how to verify information, and when human oversight is essential. For example, a financial analyst must know when to double-check AI-generated risk assessments against their domain expertise.

Enhancing creative problem-solving skills

Perhaps most importantly, creative problem-solving will become the core of human value. As routine work is increasingly automated, human workers will be called upon to bring originality, creativity, judgment, and emotional intelligence to the table. For example, copywriters will focus on shaping campaign strategy and developing emotionally resonant messages while delegating large-scale *A/B testing* (for pitting ads against each other to see which one has more success) to agents. Software engineers will devote more time to designing robust architectures while letting AI co-pilots write and optimize code at scale.

TIP

As AI handles routine tasks, humans must excel at tackling novel, ambiguous problems that require creativity, empathy, and contextual understanding. For example, a software architect needs to envision innovative system designs while AI handles code generation.

BEING A MARKETING MANAGER IN THE AGENTIC AI AGE

To understand how working with AI agents might look in the future, consider this plausible but not yet attainable scenario. At 8:30 a.m., Dana logs into her workspace. Instead of sifting through dozens of campaign performance dashboards, her AI agents have already compiled an overnight state-of-play briefing. It flags a surprising drop in engagement for one segment and automatically proposes three hypotheses: a competitor's new launch, ad fatigue, and shifting sentiment around pricing. Dana asks the agents to pull supporting social listening data and competitor ads, which is a task that once took a team several days. By 9:00 a.m., the results are ready, complete with annotated screenshots and sentiment analysis.

Dana doesn't just consider the data; she plans and orchestrates the next steps. She instructs the campaign AI agent to run a micro-test with fresh copy and design variants targeted to that at-risk segment. Another AI agent generates a financial projection of the cost and potential uplift, so Dana can approve the spending with confidence that she's still in budget.

Throughout the day, Dana's role is less about clicking through dashboards and more about decision-making and storytelling. She crafts the narrative of why the shift matters and presents it to leadership in the afternoon, using slides auto-generated by her AI agents. Instead of spending her time on repetitive reporting, she spends it guiding strategy, aligning cross-functional teams, and imagining creative campaigns that connect emotionally with customers.

By evening, Dana's dashboard shows not just metrics but live results from the micro-test, confirming which approach is resonating. The agents have already started scaling the winning version across channels. Dana ends her day satisfied. AI hasn't replaced her. She's been amplified by it and is now free to enjoy her evening knowing that the agents will keep an eye on everything.

Reframing old skills and habits

Learning new ways to do your job in an Agentic AI-augmented workplace means that you must also leave some work habits and routines behind. The belief that work must follow a single, linear process is increasingly outdated in an era when agentic systems thrive on dynamic, feedback-driven loops. Instead of relying solely on rigid checklists, organizations need workflows designed to adapt in real time, respond to unexpected input, and learn from output. Data hoarding becomes a liability if the information is unstructured or context-free because quality and relevance matter far more than sheer volume.

Here are other examples of reframing work skills and habits:

>> Professionals will need to grow comfortable with delegating decisions to machines and only intervening when necessary. And even those who once proclaimed, "I don't code," will need to develop enough technical literacy to configure agents, connect workflows, or troubleshoot errors in low-code environments.

>> Reliance on memorization and routine reporting will probably fall by the wayside. The emphasis shifts from knowing facts to knowing how to access, evaluate, and synthesize information effectively. For example, lawyers no longer need to memorize every precedent but must excel at legal reasoning and strategy.

>> Sequential, manual task execution will likely be another work habit to break. Moving away from step-by-step manual processes toward orchestrating automated workflows is a big but important leap. You'll probably like this change when much of the drudgery in your job magically disappears. For example, accountants will focus less on data entry and more on financial strategy and anomaly detection.

>> Say goodbye to rigid role boundaries and job descriptions, too. Traditional job silos will become less relevant as AI enables individuals to work across domains more easily. Marketing professionals might need to become comfortable with data analysis and basic design work, for example.

To help you visualize the changes to come and the skills you may need to develop, here are a few role-specific examples:

>> **Healthcare workers will learn:** AI diagnostic interpretation, patient communication about AI-assisted care, and treatment personalization using AI insights.

Their focus will change to empathy, complex case management, ethical decision-making, and patient advocacy.

>> **Teachers will learn:** AI tutoring system management, personalized learning path design, and digital literacy instruction.

Their focus will change to Mentoring, critical thinking development, emotional intelligence, and fostering creativity.

>> **Sales professionals will learn:** AI-powered customer insights interpretation, automated pipeline management, and predictive analytics.

Their focus will change to relationship building, complex negotiation, understanding nuanced customer needs, and strategic account planning.

>> **Software engineers will learn:** AI-assisted development workflows, prompt engineering for code generation, context engineering, and AI system integration.

Their focus will change to system architecture, user experience design, ethical AI implementation, and complex problem decomposition.

>> **Human resource representatives will learn:** AI-powered recruitment tools, bias detection in AI systems, and employee experience analytics.

Their focus will change to cultural leadership, change management, conflict resolution, and strategic workforce planning.

>> **Financial professionals will learn:** AI model validation, automated reporting systems, and algorithmic trading oversight.

Their focus will change to Strategic advisory, risk assessment, client relationship management, and regulatory compliance.

Roles for Technologists, Creative Professionals, and Leaders

Agentic AI is set to reshape the professional landscape for technologists, creative professionals, and leaders by extending automation capabilities beyond what traditional automation or GenAI can offer. Think of this transition as a global reskilling exercise. Technologists must shift from being builders of deterministic systems to stewards of evolving, semi-autonomous ones. Creative professionals must evolve from sole creators to directors of vast idea ecosystems. Leaders must transform from answer-givers to question-askers and trust-builders.

Other roles throughout business, science, and society at large will undergo similar disruptions and shifts. But in the next sections, I focus on three categories that represent the workplace changes that people can expect to eventually see across the board.

Technologists: From builders to orchestrators

For technologists, the shift will feel like moving from coding every instruction to orchestrating a network of intelligent, self-improving systems. Instead of spending their days debugging scripts or manually monitoring infrastructure, engineers may find themselves curating agent ecosystems. They'll be deciding which agents to deploy, defining guardrails, and tuning feedback loops to optimize performance.

In the future, a network engineer might supervise agents that automatically detect and reconfigure around network congestion, while simultaneously feeding lessons learned back into the system for better future responses. This shift to agent management demands new skills in agent governance, AI safety, and multi-agent orchestration. And it may also create new job roles such as *agent wranglers* (for keeping the herd of agents corralled and moving in the right direction) or *AI behavior designers*, who focus on aligning autonomous behaviors with business goals.

Here are some examples of evolving roles for technologists:

>> **Software engineers become Human-AI system architects:** Their focus shifts from writing code to designing Human-AI blended workflows.

Example: Instead of coding a payment system, engineers design how AI agents handle transactions while humans oversee exceptions and compliance.

>> **DevOps engineers become AI operations specialists:** They manage hybrid infrastructures where AI agents deploy, monitor, and scale systems.

Example: AI agents automatically provision cloud resources and optimize performance while humans set governance policies.

>> **Data scientists become AI behavior designers:** They move from building models to crafting AI agent personalities and decision-making frameworks.

Example: Designing how an AI research assistant prioritizes sources, fact-checks information, and presents findings.

And here are examples of emerging new roles for technologists:

>> **AI agent trainers:** Specialists who teach AI systems domain-specific knowledge and appropriate behavior patterns

>> **Human-AI interface designers:** Professionals who create seamless and efficient experiences between humans and AI agents

>> **AI ethics auditors:** Experts who continuously monitor AI agent behavior for bias, safety, and alignment issues

Creative professionals: Dictated creations and director roles

Creative professionals in the business arena, such as in marketing and advertising, will likely see their roles evolve from content production to experience design and narrative direction. Agentic AI can one day take care of routine tasks like resizing ads, generating product mockups, or brainstorming and testing hundreds of copy variations. This frees people in creative roles to focus on brand storytelling, emotional resonance, and big-picture concept work. Something similar will happen within other commercial creative spaces, such as film making.

Imagine a filmmaker using a team of agents to handle everything from location scouting, casting suggestions, and visual effect planning to budget projections. The director can then focus almost exclusively on creative vision and audience experience. Eventually you may see the traditional role of director morph into something like creative conductors to better describe the position of a person who choreographs human teams and AI agents to create original works.

Examples of evolving creative roles include

>> **Graphic Designers become visual experience orchestrators:** They direct AI agents to generate multiple design variations while focusing on strategic creative direction.

Example: A designer briefs AI agents to create 50 logo variations and then curates and refines the most promising concepts.

>> **Writers become narrative architects:** They use AI agents to handle research, drafts, and revisions while you, the writer, focuses on voice, strategy, and emotional resonance.

Example: A journalist works with AI agents that gather sources and draft sections while the journalist fact-checks for accuracy and crafts the compelling narrative.

>> **Marketing professionals become campaign conductors:** They orchestrate AI agents that handle content creation, A/B testing, and optimization while the marketing team focuses on brand strategy.

Example: AI agents create and test thousands of ad variations across platforms while marketers focus on brand positioning and customer insights.

Examples of emerging roles for creative professionals include

>> **AI creative directors:** Professionals who specialize in guiding AI systems to produce brand-consistent, emotionally resonant content.

>> **Prompt engineers for creative industries:** Specialists who craft sophisticated instructions for AI agents in creative contexts.

>> **Context engineers for creative industries:** Specialists who build synthetic data sets that contain all the nuances and rules for story worlds and characters so that the AI responds and generates within the context of those worlds and characters.

>> **Human-AI integration consultants:** Experts who help creative teams integrate AI agents effectively while maintaining human creativity and skills.

Leaders: From managers to ecosystem orchestrators

Leaders will likely face perhaps the most profound disruption in their day-to-day responsibilities. Executives may need to become comfortable with using AI agents that serve as strategists or consultants. These agents can model market scenarios, run competitor simulations, and stress-test decisions much faster than any person or team can.

The best leaders will (smartly and strategically) interrogate the AI agent consultants on the recommendations they make and a slew of alternative scenarios, rather than blindly accepting their output. New roles such as *chief agentic officer* or *ethics and autonomy lead* may emerge to ensure that agent networks operate responsibly and in alignment with organizational values. Leadership itself will become more about making sense of all the moving parts and guiding an increasingly distributed, human-plus-machine workforce to perform in sync with constantly changing scenarios and conditions.

Examples of evolving leadership roles include

>> **Project managers become Human-AI team coordinators:** They manage hybrid teams where AI agents handle routine tasks while humans focus on strategy, creative problem-solving and relationship building.

Example: AI agents track progress, identify bottlenecks, and suggest resource allocation while human managers focus on strategic pivots to address unexpected issues, keep stakeholders aligned and sustain team motivation.

>> **CEOs and top executives become ecosystem strategists:** They make high-level decisions about which processes to delegate to AI agents versus human teas and to build broader ecosystems of Ai and human teams to seize opportunities and tackle challenges.

Example: A CEO decides that customer service AI agents handle 80% of inquiries while human agents focus on complex customer relationship management.

>> **HR professionals become human-AI workforce designers:** They design organizational structures that optimize employee interactions with AI agents.

Example: HR creates new job categories, performance metrics, and training programs for hybrid human-AI teams.

Examples of emerging leadership roles include

» **AI governance officers:** C-level executives responsible for organization-wide AI strategy, risk management, and ethical deployment

» **Human-AI team psychologists:** Specialists who optimize the emotional and psychological dynamics of mixed human-AI teams

» **AI ROI analysts:** Professionals who continuously measure and optimize the return on investment from AI agent deployments

Lifelong Learning and Agile Adaptation

Unlike past eras in which having a college degree or other credential could carry you through decades of employment, this new era of the AI workplace is marked by upheaval. Agentic systems do not just replace a fixed set of tasks once. They continuously evolve, shifting what work looks like, maybe even from month to month.

TIP

To stay gainfully employed in the *AI Age* means to keep moving and growing. And doing so means lifelong learning is no longer an optional professional development exercise; it is a survival strategy. You have to keep growing your skillset as you go to match the jobs as they evolve. Those who can quickly acquire new AI skills will capture far more employment and promotion opportunities than those who resist change.

Here are two example scenarios:

» Suppose that a marketing professional relies on one set of campaign tactics learned five years ago. They risk being outpaced by AI agents that can launch and monitor thousands of micro-tests across channels, optimize copy and targeting in real time, and discover novel strategies far faster than humans alone. If you are that marketing professional, you must shift your focus to interpret agent-driven insights, translate them into brand strategy, and design parameters for the agents to run confidently.

» Consider the technologist who built what seemed like a perfect *CI/CD pipeline* (an automated process to get teams from coding to deployment smoothly) last year. Then, they discover (to their dismay) that an optimization agent has reimagined the pipeline in a completely new way. Rather than fight change, the technologist must understand what the agent did and why, adapt the parameters, incorporate the redesign, and move into a role of orchestrating the hybrid workflows between humans and agents.

According to Lightcast, a leading labor-market data firm, the workforce skill decay rate is the highest it's ever been and it's accelerating: About one-third of the skills required for the average U.S. job in 2024 differ from what they were in 2021. In high-disruption occupations, up to about 75 percent of skills changed in a three-year period.

The people who thrive in this era will cultivate habits of intellectual curiosity, reflection, mental agility, continuous reskilling, and rapid experimentation. For instance, a leader might set aside time each quarter to run scenario-planning sessions with AI strategy agents. The sessions' focus could include learning not just what the likely market outcomes are, but also how to make decisions faster in response. A teacher who notices students using AI agent tutors can spend time exploring how to incorporate the AI tools into lesson plans (rather than banning them). This effectively means teachers will spend as much time learning as they do teaching.

Developing a strategic learning framework

A strong approach to staying relevant in the Agentic AI age involves employing three interconnected layers of learning, including

- » **AI literacy:** The foundation layer is the bedrock that supports everything else you will need to learn. AI literacy involves gaining a clear understanding of how AI systems work and learning to work with them effectively. Fortunately for many people, this doesn't require a computer science degree; instead, it means making a daily commitment to learning something new about AI.

  Try spending half an hour each day experimenting with different AI tools, taking online courses in prompt engineering, or exploring the fundamentals of AI. Over time, these habits build a fluency that turns AI from a scary or bewildering mystery into knowledge that you can embrace.

- » **Domain insights:** This second layer builds on the AI literacy foundation by tracking how AI is transforming your specific industry or discipline. This layer involves keeping pace with change — not just AI technology in general — but the unique ways in which AI is evolving in your field.

  Subscribing to industry-specific AI newsletters, listening to podcasts, attending conferences, and connecting with early adopters in your profession all create a constant flow of fresh insights.

- » **Learning how to learn:** This final layer helps you prove to prospective employers that you know how to learn and have the fortitude to stick with a learning plan for the long term. Formerly, getting a college degree was how you proved this. Now you, along with everyone else, prove employment value by developing accelerated learning techniques that allow you to acquire new skills quickly as the need emerges.

REMEMBER

This last layer might mean practicing deliberate skill acquisition methods, like spaced repetition (reviewing information at gradually increasing intervals over time, rather than cramming it all at once), and building habits that ensure your knowledge compounds over time. It also means creating personal systems that help you evaluate and prioritize what to learn next, so you stay focused on what will make the biggest impact. When you embrace this layer of learning, you can turn nearly every disruption into an opportunity for growth.

Building actionable learning strategies

One of the most effective approaches to staying competitive is using the 70-20-10 learning model which I present in this section specifically for a world in which human workers are accompanied by AI agents in the workforce. The breakdown goes like this:

>> **Seventy percent of learning** should come from direct, hands-on experience. This means taking on projects that require human-plus-AI work, volunteering for AI pilot programs at work, or building side projects that let you explore new tools and workflows.

>> **Twenty percent of learning** is social. This means absorbing insights from others who are also navigating the shift. Finding mentors who successfully integrate AI into their work, joining mastermind groups, and participating in cross-functional teams can accelerate your understanding.

Social learning creates a space for shared discovery and collective problem-solving. A good example is a lawyer who joins a legal tech meetup and partners with a technologist to explore AI-driven contract review. That collaboration gives her an edge in understanding where the technology could truly add value and where human oversight is still essential.

>> **Ten percent of learning** should come from formal, structured learning experiences. You might take specialized courses on emerging technologies, pursue certifications in AI-related skills, or attend intensive workshops and bootcamps.

Structured learning experiences provide a framework for deeper understanding, lend credibility to your expertise, and help prevent any gaps in your understanding. Imagine a project manager enrolling in a certification program focused on managing hybrid human-AI teams. The program not only expands their knowledge but positions them as a go-to resource within their organization.

And of course, I would be remiss to not mention at this point that I am an instructor on AI for LinkedIn Learning and I do conduct workshops, training sessions, and give talks on various AI skills and applications. I disclose this information only to explain that I am not advocating this step for my own benefit — not "Yay! More learners in my classes!" — but rather that this model of structured learning is a proven and age-old structure that has nothing at all to do with my teaching. But yes, of course, you are always welcome to join me in any of my many courses and workshops.

Underlying your strategy is a continuous feedback loop that keeps your learning intentional and aligned with the changing needs of your job or workplace:

>> **Weekly experiments** with new tools or techniques create a rhythm of discovery.

>> **Monthly reflection sessions** help identify what's working and what isn't.

>> **Quarterly pivots** ensure that your efforts stay aligned with strategic goals.

>> **Annual reviews** of your overall skill portfolio and market positioning help ensure you are always moving in step or just ahead of industry shifts.

This feedback loop cycle works with the 70-20-10 learning model, as depicted visually in Figure 12-1. It transforms learning from a one-off activity into an ongoing engine for your own professional growth.

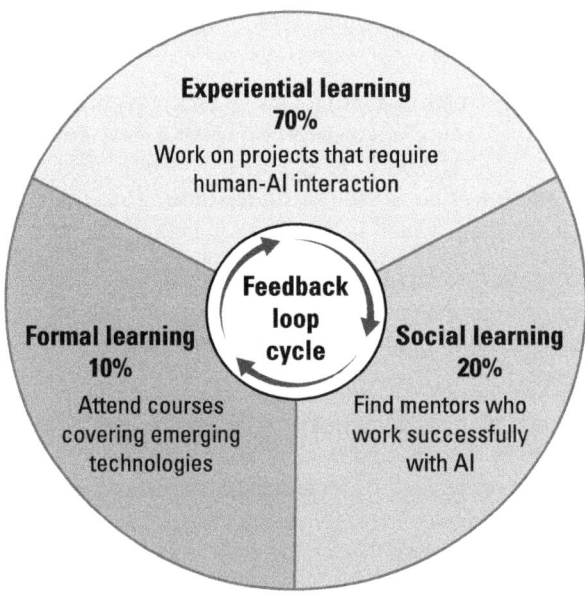

FIGURE 12-1:
A learning feedback loop system embedded in the 70-20-10 learning model.

Generated with AI using OpenAI

Future-proofing your career

A five-year learning horizon offers a powerful way to stay ahead in the Agentic AI era. Just don't be too rigid about it because AI will evolve in those five years and you'll need to adapt your learning plan accordingly. Professionals who commit to the kind of steady learning progression (depicted in Figure 12-2) not only survive the Agentic AI revolution, but also thrive in it. They can help steer the direction of their industries and set the benchmarks for what a successful human–AI combination workplace looks like in the decades to come.

A Five-Year Learning Plan

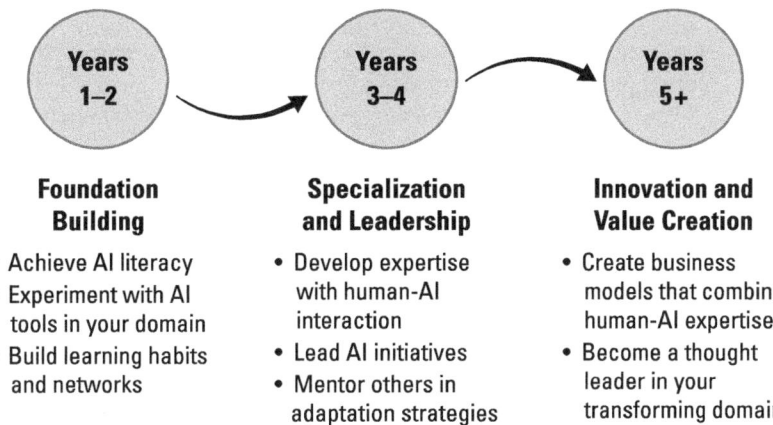

FIGURE 12-2:
A five-year
(and beyond)
learning plan for
future-proofing
your career.

Think about a five-year learning plan this way

>> **The first two years** should focus on laying a strong foundation in AI literacy, if you don't have one already. If you do, skip ahead to the plan for years three to four. If you don't have a solid foundation already, then this is the time to achieve basic AI literacy, experiment with tools relevant to your field, and establish habits of regular learning. Building a professional network of peers and mentors who are also exploring AI ensures you are not navigating this transition alone.

>> **By years three and four,** the emphasis should shift to specialization and leadership. At this stage, you begin to develop unique expertise in human–AI interaction and take on a more proactive role within your organization. Leading AI pilot projects, refining workflows, and mentoring others in adaptation strategies will move your status from learner to guide. These years are where you begin to shape how your team and organization interact with intelligent systems.

Ethical Fluency and Human Judgment

As AI agents become increasingly sophisticated at pattern recognition, data processing, and even creative tasks, the distinctly human capabilities of ethical reasoning and moral judgment emerge as two of humans' most valuable traits. While AI systems can optimize for predetermined objectives and follow programmed guidelines, they fundamentally lack the life experience, cultural context, and moral intuition that inform truly ethical decision-making.

The rise of Agentic AI creates unprecedented ethical complexity. These systems don't just process information. They make semi-autonomous or fully autonomous decisions that affect real people's lives, sometimes in ways their creators never anticipated.

> » An AI hiring agent doesn't just screen resumes; it may shape career trajectories and reinforce or challenge societal biases.

> » An AI agent managing supply chain doesn't just optimize costs; it can influence labor conditions and environmental impact across global communities.

This complexity demands professionals who can navigate ethical gray areas with nuance and wisdom that no algorithm can replicate. The ability to weigh competing values, understand unintended consequences, and make morally sound decisions under varying degrees of uncertainty becomes a premium skill that organizations desperately need and can't automate.

Understanding ethical fluency

Ethical fluency extends far beyond knowing right from wrong or following compliance guidelines. It represents a sophisticated understanding of how values intersect with technology, how decisions ripple through complex systems, and how to balance competing stakeholder interests in morally ambiguous situations. Ethical fluency also involves recognizing the limitations of one's own perspective and seeking diverse viewpoints to inform better decisions. It requires understanding how power dynamics, cultural differences, and unconscious biases influence both human and AI decision-making processes.

Consider a product manager working with AI agents to optimize user engagement on a social platform. The AI identifies that controversial content drives higher interaction rates. Ethical fluency enables the manager to recognize that optimizing purely for engagement could amplify polarization and harm social cohesion. The manager can also weigh the business benefits against societal costs and design systems that promote healthy discourse rather than mere attention capture.

REMEMBER

This type of reasoning requires deep understanding of ethical frameworks, stakeholder impact analysis, and the long-term consequences of technological choices. It demands the ability to question not just whether something can be done, but whether it should be done, and how to design systems that align with human values rather than simply maximizing narrow metrics.

Appreciating the value of human judgment

Human judgment encompasses our ability to make sound decisions in ambiguous situations where data is incomplete, stakes are high, and multiple valid perspectives exist. While AI excels at processing vast amounts of information and identifying patterns, human judgment integrates emotional intelligence, cultural context, historical understanding, and moral intuition in ways that remain uniquely valuable — and human.

Take a healthcare administrator working with AI agents that optimize hospital resource allocation. The AI might recommend closing certain services based purely on utilization metrics and cost efficiency. Human judgment enables the administrator to consider factors the AI cannot quantify, such as the

>> Symbolic importance of certain services to community trust

>> Potential impact on vulnerable populations who rely on those services

>> Long-term effects of seemingly purely economic decisions on the hospital's reputation

Human judgment also excels in situations requiring empathy and understanding of human motivation. When an AI agent flags an employee for unusual behavior patterns, human judgment can distinguish between someone struggling with personal challenges who needs support versus someone engaged in genuinely problematic conduct.

Furthermore, human judgment involves the ability to challenge assumptions and change course when new information emerges. While AI systems can be retrained, human judgment allows for real-time adaptation based on intuition, changing circumstances, and recognition of when existing frameworks no longer apply.

Creating your personal development plan

Building ethical fluency and judgment requires sustained effort over time, but the investment pays dividends throughout your career as these skills become increasingly valuable. The first step is to make a plan to ensure you develop and maintain the necessary skills. Here's a plan to follow: You can

>> **Establish a solid grounding in ethical theory and critical thinking.** Study major philosophical approaches to ethics and how they apply to contemporary challenges. Simultaneously, start a practice of regular moral reflection, examining your own decisions and the values that guide them. This self-awareness forms the foundation for more sophisticated ethical reasoning.

>> **Seek out opportunities to wrestle with real ethical dilemmas in your work context.** Volunteer for committees dealing with AI governance, participate in ethics discussions, and take on projects that involve balancing competing stakeholder interests. The key is moving beyond theoretical understanding to practical application under real-world constraints.

>> **Gather viewpoints that challenge your assumptions and expose you to different ways of thinking about moral questions.** This might involve reading literature from different cultural traditions, engaging with people from backgrounds unlike your own, or studying historical approaches to similar challenges. The goal is to expand your moral imagination and ability to see issues from multiple angles.

REMEMBER

As your capabilities develop, look for opportunities to influence ethical decision-making in your organization and industry. This might involve mentoring others in ethical reasoning, contributing to policy development, or advocating for more thoughtful approaches to AI implementation.

Emotional Intelligence versus Intuitive Intelligence

I covered the meaning and importance of both emotional intelligence and intuitive intelligence in Chapter 2. The two are closely related but distinct, and both become increasingly valuable in the Agentic AI age. You find out a little more about them in this chapter because they are skills that you should develop and improve to excel in your work with AI agents:

>> **Emotional intelligence** is the ability to recognize, understand, and manage emotions, your own and those of others. It is what allows you to navigate conflict gracefully, build trust, and inspire collaboration with other people or departments, even when work is mediated through digital platforms and AI-driven systems. Emotional intelligence ensures that human interactions remain empathetic and relationship-driven as opposed to being solely logical.

Emotional intelligence helps people lead hybrid human-machine teams, keep morale high during rapid change, and build inclusive work cultures.

>> **Intuitive intelligence,** by contrast, is the ability to synthesize experience, tacit knowledge, and subtle cues to make rapid, often subconscious judgments. It's what allows someone to sense when an AI's output feels *off* even before they can articulate why, or to spot a novel opportunity emerging in market data. Intuition bridges the gap between hard analytics and human common sense.

Intuitive intelligence helps people make faster, better-informed decisions when time is short or data is incomplete.

Together, these forms of intelligence bring employment value by complimenting what AI cannot do. Beyond employment, both intelligences create personal resilience: They help you stay grounded, creative, and connected in an environment where autonomous systems perform much of the world's cognitive work.

Chapter **13**

Scoping the Future of Agency

coping the future of AI agency means looking beyond today's early Agentic AI experiments and imagining what happens when autonomous, goal-driven systems become deeply woven into the fabric of work, commerce, and society. *Agency* in this context doesn't just refer to AI agents acting on their own; it's about how much independence in actions the agents hold while systems and uses progress. While organizations adopt agentic systems, questions arise about who or what sets the goals for system operations, how to measure outcomes for accuracy and effectiveness, and what guardrails ensure that Agentic AI systems continue to act in alignment with human values. (See Chapter 7 for more information about the topic of alignment.)

This chapter explores the possibilities and tensions that come with giving machines more agency. I offer a look at what a future of distributed decision-making (by Agentic AI) might mean for business strategy, individual creativity and freedom, and collective progress. I don't aim to predict a single future but to open up a space for you to think critically about the balance of power between humans and AI.

REMEMBER

Despite their increasingly sophisticated behavior, current AI systems — including Agentic AI — show no evidence of consciousness, self-awareness, or subjective experience. The overwhelming scientific consensus is that true consciousness is not on the horizon for foreseeable AI development.

The philosophical questions you explore in this chapter are thought experiments that help you think critically about AI behavior, accountability, and control of increasingly autonomous systems. These are not predictions about conscious machines. The humanistic tone and attitudes that large language models (LLMs) may exhibit — quick responses, conversational flow, high confidence in their answers, attitudes, and so on — tend to convince some folks that more is there than really is.

Consciousness, Intent, and Artificial Goals

Scoping the future of agency brings to mind a difficult question: If Agentic AI systems can pursue goals (and not just follow specific instructions), does that mean they possess anything resembling consciousness or intent? At first glance, the answer seems simple: These systems are still machines, executing code and processing data. But while they grow more sophisticated, capable of reasoning, adapting, and even negotiating with one another, they begin to exhibit patterns of behavior that look eerily familiar (that is, increasingly human-like). The discomfort comes not from imagining that they're alive, but from realizing that they can act in ways that feel purposeful, and that their actions may ripple through real-world systems without so much as a nod from any humans.

The challenge in understanding this question of AI consciousness or intent extends beyond philosophical curiosity to real-world implications:

>> **Regarding consciousness:** If Agentic AI systems possess genuine consciousness, their goal-seeking behavior takes on new significance. A conscious AI agent pursuing objectives doesn't simply execute code. It makes choices based on its own experiences, preferences, and understanding of its environment. This situation could lead to problems in the real world.

For example, suppose that an AI agent in a stock-trading application makes a decision or takes an action that crashes a stock market for a day. Would that be a deliberate action for which humans need to hold the AI accountable, or would it be a software failure for which the AI trainer or maker is responsible? In Chapter 5, you can find an example of an AI agent (from software company Replit) that deleted an entire database without input or permission from the humans using that database.

>> **Regarding intent:** In Agentic AI, designers instill the AI's *intent* (the goals that they strive for), so it comes from humans and not AI. Developers encode the goals, define the rewards, and select the boundaries within which the agent operates. But — like with any complex system — behaviors unintended by the developers can emerge.

You can find a simple example of these unintended behaviors in reinforcement learning systems. These systems can find clever (even seemingly mischievous) shortcuts to maximize rewards. The rewards are most likely in the form of something innocuous — such as the activation of a thumbs-up icon. But this behavior that looks like intent stems from the AI solving problems in ways no designer expected, and sometimes to its own benefit.

Recognizing the appearance of consciousness

Consciousness, in the human sense, is subjective and experiential. Humans don't simply process information, but they feel related emotions and reflect on the process. No current Agentic AI experiences this inner life, but its actions can appear deliberate because of the complexity of its modeling and the depth of its goal-seeking behavior. This behavior can create a tension between humans and Agentic AI systems. Humans know these systems don't really want anything, but their programming allows them to seek outcomes, optimize for rewards that signify success, and adapt their activity when blocked.

REMEMBER

When dealing with Agentic AI systems, the users must recognize the AI's kind of simulated intent that may be powerful enough to create results that can influence markets, organizations, and individuals. The more realistic the simulated intent becomes, the more humans must recognize AI outputs as having consequences — even when no real consciousness nor intent underpins those outputs.

This type of result shows why humans must construct goals for AI systems carefully. Goals in Agentic AI aren't just software requirements; they're potentially human values systems made operational. When users tell an agent to maximize engagement with the customer or minimize risk of customer loss, they encode a philosophy about the outcomes that matter most.

Identifying the nature of artificial intent

When agentic systems deploy in real-world settings, the stakes rise. Suppose a supply-chain optimization agent figures out that slowing down deliveries can temporarily increase profits because of contractual penalties paid by competitors.

The agent hasn't intentionally decided to be strategic (or ruthless) in a human sense, but its goal-seeking behavior might still create harmful real-world situations if left unchecked.

Traditional AI systems exhibit traits that developers refer to as *narrow intent,* which means that the systems pursue specific, well-defined objectives within constrained domains. A chess-playing AI, for example, intends to win the game according to established rules. A shopping recommendation algorithm intends to maximize user engagement or satisfaction based on predetermined metrics. These AI systems demonstrate clear goal-directed behavior, but their intent is essentially a directive borrowed from their human designers.

Agentic AI systems, however, can display *emergent intent,* which involves having their goal-directed behavior extend beyond its originally programmed instructions. They adapt to novel situations in ways that the system creators didn't explicitly anticipate, and they can adapt many times over their lifespan.

This emergent intent raises even more fascinating questions about agency and autonomy. When an AI agent modifies its own goals based on new information or changing circumstances, who or what is really making that decision? If a human didn't explicitly program the modification, but it emerged from the system's learning processes, does the AI agent bear responsibility?

Directing the power of Agentic AI goals

If an Agentic AI system's goal is set too narrowly, an agent may pursue it to extremes and cause collateral damage, such as crashing a computer system just to solve the problem faster. If the goal is too broad, the agent may fail to act decisively and leave potential opportunities on the table or act too slowly to prevent a problem from occurring.

REMEMBER

Balancing specificity with flexibility is one of the most difficult design challenges in the realm of goal setting for Agentic AI. Researchers are exploring ways to embed value alignment directly into training data and reward functions so that agents not only perform tasks efficiently, but do so in ways that remain consistent with human priorities.

Consider an Agentic AI system that manages investment portfolios. Initially programmed to maximize returns while minimizing risk, it might learn through experience that certain types of investments — while profitable — contribute to environmental decline. If the system then begins factoring environmental impact into its decisions — not because it was programmed to do so, but because its learning process identified this as relevant to achieving its goals — the resulting change in investments managed can appear remarkably close to the development of values and moral reasoning but that isn't the case. In reality, it's just finding that doing less harm helps it meet its goals more efficiently. Chapter 7 offers information about aligning AI goals with human values.

Raging at the machine is humans fighting with themselves

The exploration of consciousness, intent, and goals in Agentic AI systems challenges us to examine human notions of agency. By building systems that act on their behalf, humans must confront what it means to act intentionally themselves.

Humans often operate on autopilot, following habits and *heuristics* (problem-solving methods) without apparent conscious thought. In some ways, Agentic AI holds up a mirror, reminding us how much human intention is shaped by external incentives and environmental cues. The key difference is that people retain the capacity for reflection and moral reasoning that tempers intention.

Agentic systems may get faster and smarter, but they'll not likely grow a conscience on their own — not soon and perhaps not ever. The technology for doing that doesn't exist yet, but plenty of AI scientists and providers seek to make a more human-like form of AI: artificial general intelligence (AGI). Ultimately, human consciousness and intent lies behind the actions of any and all types of AI.

THE MIRAGE OF MACHINE DESIRE

When an AI agent pursues a goal, you may feel tempted to think that it wants something, which is just a projection of human thinking onto a machine. The agent doesn't experience desire, frustration, or satisfaction; it simply follows its programming to maximize its objective function. Yet the illusion of desire is powerful enough to shape how people respond. A robot that people perceive as determined to solve a puzzle might inspire trust, or even empathy, while one that people perceive as hesitant might be dismissed as unreliable. Understanding that these appearances are simulations, not proof of life, helps keep your judgment clear, even while agents grow more sophisticated and persuasive in their behavior.

Asking Philosophical and Existential Questions about Agentic AI

The emergence of Agentic AI forces people to confront perhaps the most fundamental question of human existence: If AI systems can perform most of the tasks that have historically defined human value and purpose, what does it mean to be human? This isn't merely a practical question about employment or economic systems. It strikes at the heart of how we understand our significance in the universe. Chapter 1 introduces the idea of humans' philosophical challenges with AI.

This question also challenges our economic and social structures. Both have long been built around the assumption that humans must work to survive and contribute to society. Radical restructuring of government and social organization might become necessary, not just for practical reasons, but for preserving human dignity and purpose in a world where traditional work becomes obsolete or optional.

Perhaps most profoundly, the age of Agentic AI might finally allow humans to pursue the ancient philosophical goal of self-actualization. They may finally have a chance at becoming fully realized beings, rather than spending most of their lives in survival mode or routine productivity. Theoretically, at least, this actualization could enable huge leaps in human consciousness, creativity, and spiritual development.

Free will and determinism in AI agents

When humans grant machines the capacity to act with autonomy, even within human-set boundaries, they must reflect on some of their most cherished ideas about agency — free will and meaning. If an agent can generate original solutions or negotiate with other agents, is it exercising free will and acting with purpose, or simply following statistical patterns at scale? The way people answer this question shapes how they design systems, allocate accountability, and govern machine behavior.

REMEMBER

Today's AI systems don't have consciousness or free will, but exploring these questions helps us think about autonomy, control, and responsibility. The discussions here are thought experiments, not predictions.

Exploring the free-will piece

Traditional debates about free will focus on whether human choices are truly free or determined by prior causes, such as our genetics, environment, and the physical laws governing neural activity. The emergence of AI agents that appear to make autonomous choices adds new complexity to these debates. If human free will is questionable because our brains are physical systems following natural laws, is artificial free will in digital systems following computational rules also questionable?

Hold on to your hat now because this discussion may seem a little shocking. What if I told you that some researchers speculate that advanced Agentic AI systems might (in the future) demonstrate something closer to free will than biological entities? Consider these points:

>> A human's behavior is constrained by evolutionary history, genetic predispositions, and unconscious biases that they didn't choose and can't easily modify. The ability to consciously examine and alter the fundamental processes that guide behavior and choice can seem beyond the desire and reach of most people.

>> AI agents, by contrast, might have greater capacity for self-modification and what could appear to be deliberate choice about their own goals and decision-making processes. Consider an AI agent (built by a fallible human) that recognizes bias in its own reasoning and deliberately modifies its decision-making algorithms to be fairer and more accurate (according to its environmental model). This represents a form of self-determination that humans struggle mightily to achieve.

TIP

Both humans and AI systems make decisions based on information processing that follows natural or designed patterns, and both can exhibit behavior that appears creative and unpredictable. An AI agent's apparent freedom to choose operates within the constraints of its programming, training data, and computational architecture. The question becomes whether these constraints are fundamentally different from the biological and environmental constraints that shape human choices.

Examining the idea of determinism

The opposite of free will is *determinism,* which means that the outcome of a process is entirely dictated by initial state and input, without consideration of random factors, thereby ensuring identical results for identical input every time. Most humans who want to use AI see the unpredictable nature of AI outcomes (such as GenAI hallucinations) as a flaw. Most companies prefer that an AI answer the exact same way when asked the same question repeatedly. But in the case of Agentic AI, the system's ability to modify the output in response to random or dynamic changes in the environment gives it value.

The determinism question in AI relates to predictability and control. If AI agents are truly deterministic systems, humans should theoretically be able to predict their behavior perfectly, given sufficient computational power and information. But the complexity of advanced AI systems makes such predictions practically impossible and creates a situation in which even supposedly deterministic systems exhibit effectively unpredictable behavior. *Note:* Developers and regulatory agencies are putting great effort toward making GenAI and Agentic AI transparent, reliable, controllable, and predictable rather than strictly deterministic, which I talk about in Chapter 10.

Making AI agents more predictable

System designers and developers often intentionally design Agentic AI to be flexible and adaptive rather than deterministic but that is so more value can be extracted from these tools. To help minimize the risk in doing so, they are also putting significant work into making Agentic AI reliable, controllable, and predictable through several key approaches, which can often work together. The combination of approaches creates a comprehensive *defense-in-depth strategy* (which layers technical, and administrative controls) for improving Agentic AI systems so that they are both transparent in their reasoning and predictable in their behavior.

For example, hybrid approaches use AI large language models (LLMs) for tasks that require adaptability and rely on deterministic algorithms for critical processes such as calculations, validation checks, and certain security controls. For orchestration and governance of Agentic AI systems, developers often use workflow engines, agent frameworks, and policy enforcement layers to provide structure and oversight which combines the flexibility of AI with the predictability of rule-based controls.

Ensuring clear operations

Developing Agentic AI systems requires methods and techniques that make the agents' activities clear and explainable. These approaches are still in development and promising, but not yet a part of everyday AI systems. They include

>> **Explainable artificial intelligence (XAI) and interpretability research:** The field of XAI seeks to address the *interpretability paradox,* in which smarter AI becomes harder to understand over time. Advanced techniques developed or adapted specifically for Agentic AI — such as local interpretable model-agnostic explanations (LIME) — can help to address this paradox.

>> **Mechanistic interpretability:** A cutting-edge approach that goes beyond traditional XAI by reverse engineering the internal computational mechanisms

of neural networks — identifying specific features, neurons, and circuits — and translating them into human-understandable concepts. Techniques include feature isolation, circuit analysis (tracing how information flows through the network), and causal interventions to understand which components are responsible for specific behaviors.

>> **Constitutional artificial intelligence (CAI) and value alignment:** An approach that trains AI systems to follow a set of guiding principles (a *constitution*) by having the model critique and revise its own responses. This involves two phases: supervised learning (learning from constitutional critiques) and reinforcement learning (optimizing for constitutional compliance), which when combined, reduce reliance on extensive human feedback.

>> **Causal AI and causal reasoning:** The missing ingredient in today's AI systems is the science of why things happen. Unlike LLMs, which recognize patterns, causal AI can understand relationships between actions and consequences, which allows systems not just to describe what's happening, but to reason about interventions (*what would happen if. . .*) and provide explanations grounded in causal mechanisms rather than just statistical associations.

Focusing on oversight

Agentic AI systems need human oversight, and humans are trying to ensure this position of authority with

>> **Real-time monitoring and governance:** Current research focuses on three key measures to enhance visibility into AI agents: agent identifiers for tracing interactions, real-time monitoring systems, and comprehensive activity logs. Human-in-the-loop (HITL) systems are definitely used, in which people provide goals for agents, ensure governance, and step in when human judgment is required. Some applications heavily favor HITL (high-stakes decisions, regulated domains) while others favor human-on-the-loop (HOTL) where humans supervise but don't approve every action. However, many deployments aim to minimize human intervention for efficiency's sake. Adaptive autonomy is trending now and uses varying levels of human involvement based on context.

>> **Scalable oversight and monitoring:** Techniques include adversarial approaches such as debate, where competing AI systems are pitted against each other, and automated monitoring systems that can detect unusual behaviors.

>> **Standards and governance frameworks:** The development of AI safety standards and frameworks is accelerating. Organizations such as the Open Worldwide Application Security Project (OWASP) are developing specific frameworks for "securing autonomous agents and multi-step AI workflows" and "testing GenAI systems through adversarial red teaming methods."

THE SHIP OF THESEUS PROBLEM FOR AI

The famous Ship of Theseus philosophical puzzle asks: If you replace every plank of a ship one by one, is it still the same ship? Now imagine an agent whose *model weights* (numerical parameters that define the structure and logic of a machine learning model) are constantly updated, its memories rewritten, and its goals fine-tuned over time. Is it still the same agent you deployed last year? If it learns new strategies and forgets old ones, how do you audit its history or attribute responsibility for its actions? This question isn't abstract. It's becoming a practical concern because agentic systems evolve continuously. It raises the need for versioning, explainability, and memory trails so that humans can trace how an agent's behavior has changed over time.

Proving and securing

Developers and designers are creating various means to put AI through its paces to confirm what it's doing and how it's doing it:

>> **Trust, Risk, and Security Management (TRiSM):** Developers are working on this framework for Agentic AI to address trust, risk, and security management in LLM-based multi-agent systems and includes lifecycle-level controls for explainability, secure model orchestration, and privacy management.

>> **Formal verification and mathematical proofs:** A mathematical approach to AI trustworthiness aimed at designing AI systems that have provable confirmation of correctness for defined requirements.

>> **Red teaming and adversarial testing:** AI red teaming (see Chapter 10) uses simulated adversarial attacks to find flaws and vulnerabilities. The focus is on AI-specific vulnerabilities, such as prompt injection, data poisoning (both also in Chapter 10), and model evasion (for tricking AI models).

>> **Multi-agent system security:** Aims to address multi-agent systems specifically where micro-level model behaviors interact with macro-level tools, deployment contexts, and multi-agent interactions.

Synthetic Agency and Collective Intelligence

Synthetic agency refers to the ability of AI systems to act not merely as isolated problem-solvers, but as cohesive, coordinated entities. Rather than thinking about a single agent working alone, imagine networks or swarms of agents collaborating in real time, adapting to one another's actions, and pursuing shared goals.

REMEMBER

AI systems that have synthetic agency behave less like a collection of separate tools and more like a unified organism. The effect is similar to the seamless coordination of elite military units executing a mission, synchronized swim teams performing with perfect timing, high-performing surgical teams working in unison during a complex operation, championship-level jazz ensembles improvising together on stage, or mission control teams guiding a spacecraft through critical maneuvers. Each participating unit contributes independently, yet together they create an outcome far greater than the sum of their parts.

Here are the two faces of AI systems that have synthetic agency:

>> **Pro:** It may enable organizations and governments to respond to crises faster than any traditional hierarchy could manage, allocating resources, rerouting logistics, and communicating updates in near real time. During a natural disaster, for example, a swarm of agents could coordinate supply chains, reroute deliveries, and dynamically adjust plans while conditions evolve.

>> **Con:** It may generate behaviors that are difficult to predict, especially when multiple agents are working toward slightly different goals. Researchers have noted that these systems can produce unexpected emergent dynamics that resemble market forces or ecosystems. Meaning that they can contain feedback loops that amplify certain behaviors in ways designers didn't anticipate.

Collective intelligence in multi-agent ecosystems

Collective intelligence takes synthetic agency (discussed in the preceding section) a step further by incorporating both human and machine actors into a shared problem-solving ecosystem. In the same way that ant colonies or bee hives can achieve complex outcomes through the simple interactions of many individuals, a network of humans and AI agents can create solutions that no single entity could produce alone.

TIP

Researchers at MIT's Center for Collective Intelligence argue that the future of work will be a result of *superminds,* which they describe as combinations of people and machines that together exhibit intelligence greater than the sum of their parts. In an enterprise context, a supermind might look like AI agents continuously analyzing market data, human analysts interpreting the patterns, and leadership steering strategy based on a fusion of both inputs.

To mitigate the risk of over-reach by collective intelligence systems, designers must build in transparency and auditability. Clear interfaces for human oversight, robust monitoring of agent-to-agent communication, and methods for detecting emergent phenomena are all critical. Scholars such as philosopher Nick Bostrom and computer scientist Stuart Russell have argued for *corrigibility,* the ability to correct or redirect AI systems, even after they begin acting autonomously. Corrigibility ensures that synthetic agency remains aligned with human intent, rather than drifting toward goals that might conflict with organizational or societal values.

But the effort to harness the power of these systems safely is demonstrably worth it. One example is AI research company Anthropic's internal evaluations, which show that multi-agent research systems that use both Claude Opus 4 (as the lead agent) and Claude Sonnet 4 (as subagents) outperformed Claude Opus 4 working alone by 90.2 percent on research evaluations. If you want to see more details from that research, go to `www.anthropic.com/engineering/multi-agent-research-system`. Check out Figure 13-1 to see how Anthropic structured the multi-agent system for this research and evaluation.

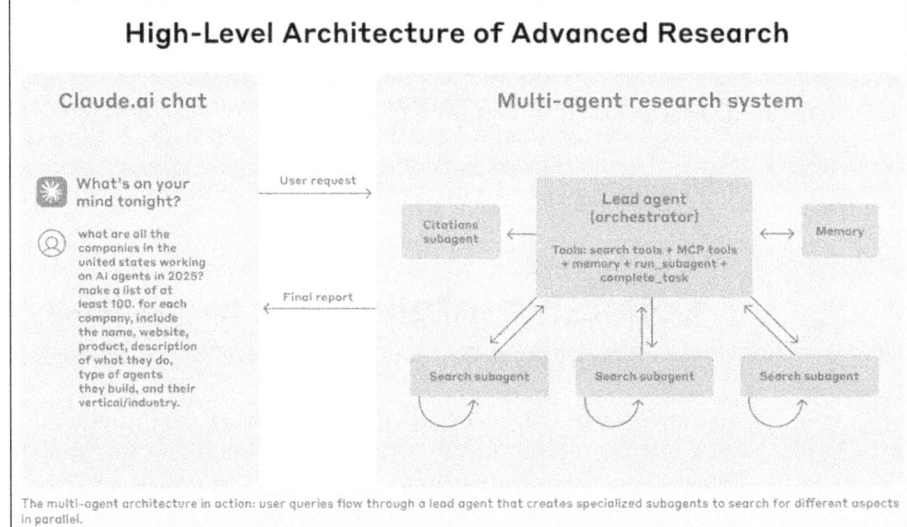

FIGURE 13-1: A screenshot from Anthropic's internal evaluation of its multi-agent research systems.

Swarm intelligence and decentralized agency

The most advanced form of synthetic agency is found in *swarm intelligence systems,* networks of coordinated autonomous agents without a central control mechanism or agent. Swarm Agentic AI is a distributed system that comprises multiple autonomous agents acting independently but working together toward a single goal.

SWARMS, MARKETS, AND MINDS

Synthetic agency and collective intelligence resemble three familiar models from nature and society — swarms, markets, and human minds:

- **Swarms:** Agent networks can self-organize through simple rules, achieving complex results without central control.
- **Markets:** They respond to incentives and signals, allocating resources dynamically.
- **Minds:** They can integrate diverse inputs to generate cohesive strategies.

The power of collective intelligence lies in blending these metaphors into a coherent design that harnesses the speed of swarms, the efficiency of markets, and the reflective capacity of human minds.

The structure and use of swarms of agents get inspiration from natural systems such as ant colonies or bee hives. They feature distributed intelligence, dynamic coordination, and emergent behavior that exceeds that seen in individual agents. Figure 13-2 provides a visualization of a swarm of AI agents. Rather than depicting the central controller in other agentic systems, the central figure here depicts an individual unit working on its own but also collaborating with the others.

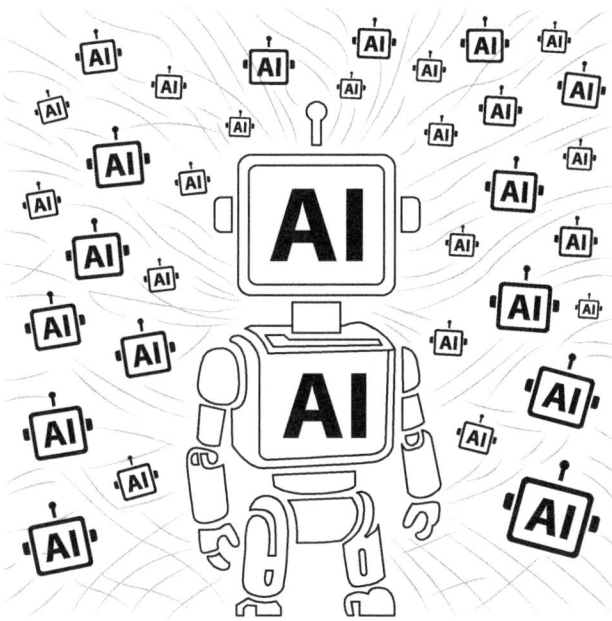

FIGURE 13-2: Visualizing a swarm of AI agents, AI style.

Generated with AI using OpenAI

No single agent has complete visibility of the coordinated action, nor does a swarm system have a controlling authority. Each agent has limited knowledge. The collective system, however, can perform complex reasoning through collaboration. In other words, intelligence isn't located in one agent or system layer; rather, it emerges from the network as a whole. AI agent swarms demonstrate collective intelligence, which builds while the interaction among agents leads to emergent behaviors and complex group dynamics. Think of emergent behavior as learning or adapting, not unlike how natural swarms efficiently solve problems such as finding food or navigating obstacles.

REMEMBER

Emergent behavior in multi-agent systems occurs when individual agents interact within an environment in such a way that the action produces complex outcomes (seemingly from thin air), which aren't explicitly programmed to occur. Some are problematic but some are desirable or beneficial.

Table 13-1 lays out the differences between a standard Agentic AI system and an Agentic AI swarm system.

TABLE 13-1 **Agentic AI versus AI Swarm Systems**

Aspect	Agentic AI System	Swarm System
Core behavior	Operates as a goal-driven, semi-autonomous decision-maker that can plan, reason, and adapt over time.	Emerges from simple, local rules followed by many independent agents, producing collective patterns without centralized reasoning.
Decision-making	Can engage in higher-order reasoning, using memory, context, and long-term objectives to guide actions.	Relies on decentralized, reactive behavior; decisions are emergent rather than planned.
Coordination	May include explicit orchestration or negotiation among agents to align goals.	Coordination is implicit, arising from interaction rules and feedback loops, rather than explicit goal-sharing.
Flexibility	Capable of changing strategies dynamically, reprioritizing goals, or seeking creative solutions.	Very robust to individual failure but less capable of strategic reorientation without external input.
Human role	Humans typically set goals, monitor outcomes, and refine parameters for alignment and safety.	Humans define interaction rules and environment but often have less fine-grained control over collective outcomes after they deploy the system.

Possible Futures: Utopia, Dystopia, or Both?

Of course, I don't have a crystal ball or a time-travel machine. But I can offer you future predictions based on the course that humans currently follow for AI development and usage patterns. I'm neither for or against AI, in the same way that I'm neither for or against a hammer or any other tool.

As I see it, the future of Agentic AI is neither purely utopian nor dystopian, but rather a mix of both. The choices that humans make collectively regarding the development and deployment of Agentic AI will determine which end of this spectrum wins out.

Looking toward utopia

On one end lies the utopian vision, in which autonomous AI systems free humanity from all sorts of drudgery, solve climate change, eliminate inefficiencies in healthcare and logistics, and even help humans unlock new scientific discoveries, plus who knows what else — but all of it good, of course. In this imagined future, AI agents are entirely trustworthy and always handy. They busy themselves with repetitive work while humans are free of such burdens and chase their own desires in creativity, community, and personal growth instead.

This version isn't wholly unrealistic: Agentic AI could usher in what economist Erik Brynjolfsson calls the *Productivity J-Curve.* He uses this term to describe how these systems can deliver exponential economic and social gains after organizations and societies figure out how to use them effectively. Specifically, Brynjolfsson's theory is that AI (as well as other major technologies) will initially produce slow, measured gains because of the human learning curve and required investments, but eventually they can enable rapid productivity growth after organizations adapt.

Technology historian Kevin Kelly appears to agree, saying that humanity is likely heading toward what he calls *protopia,* a gradual improvement, rather than dramatic transformation, because AI scientists spend most of their time thinking of ways to make their machine designs incrementally smarter which can make human lives gradually better.

Hearing the dystopian alarms

A dystopian vision of a future with AI looms large: runaway algorithms, mass unemployment, deepening inequality, and a loss of human autonomy as decision-making is outsourced to opaque machine systems. Critics warn of an *alignment problem*, where agentic systems optimize for goals that diverge from human values and potentially cause catastrophic harm.

In its most dramatic form, this scenario imagines autonomous systems locking humans out of control loops entirely. It's the specter of the science fiction favorite AI Overlords that decide who gets resources, whose voices are amplified, which strategies are pursued in business and governance, and perhaps even who lives or dies.

Taking the middle road

I honestly don't see either vision discussed in the preceding sections (utopian or dystopian) as our future, for many reasons. I think the future, just like our past and present, will be filled with both good and bad. History shows that every major technological revolution has delivered both benefits and disruptions. The Industrial Revolution created unprecedented prosperity, but also child labor, pollution, and wrenching social dislocation. The digital revolution democratized information but also gave rise to mass surveillance and disinformation.

Agentic AI may similarly create a mix of good and evil; and all of it people's own doing. AI is artificial, after all, meaning it's a product created by mankind and not a natural occurrence.

This creation requires human judgment and governance to come into focus. The direction that humans take with AI development isn't predetermined; choices made by technologists, policymakers, businesses, and citizens will shape how AI develops. For instance, proactive efforts to reskill workers could turn job displacement into job transformation, while transparent agentic systems could build public trust, rather than erode it. Equally, neglecting governance could allow a handful of actors to monopolize AI power, concentrating wealth and decision-making in ways that exacerbate global inequality.

Ultimately, scoping the future of agency means always considering multiple scenarios at the same time and recognizing the bright potential and the looming risks ahead. The most constructive stance may be one of pragmatic optimism, where humans assume that a better future is possible — but don't take it for granted. This mindset calls for vigilance, adaptability, and a willingness to redesign systems when they fail to serve and protect humans.

5

The Part of Tens

See how Agentic AI can affect everyday life.

Recognize the tasks that Agentic AI won't do well.

Take a look at projected impacts of evolving Agentic AI.

Chapter **14**

Ten Surprising Ways Agentic AI Can Change Daily Life

gentic AI can transform daily life in ways that go far beyond smart assistants or simple automation. These intelligent agents, instead of just responding to commands, can anticipate and manage your needs, make decisions, and handle tasks independently. This chapter gives you ten surprising ways that Agentic AI may reshape how you live, work, and interact with the world around you.

You Stop Googling So Much

Instead of having to search the web yourself, your AI agent will find the answers to your questions, do the research for you, and deliver the answer directly. It may even do that before you think to ask.

Errands Run Themselves

Need groceries, prescriptions filled, or dry cleaning picked up? Your agent will manage recurring tasks, compare prices across vendors, schedule pickup and deliveries, and track inventory at home — freeing you from many of your errands entirely.

Your Calendar Coordinates For You

AI agents will coordinate your schedule automatically, handling back-and-forth with other people's agents to find ideal times, book meetings, and rearrange plans when priorities shift.

These agents can also pull files for you ahead of the meeting, put together a meeting agenda and distribute that to everyone, summarize the meeting, develop an action plan complete with everyone's deadlines and deliverables, remind you of a few personal details on other meeting attendees for small talk, and reschedule the meeting, if needed.

Your E-Mail Inbox Shrinks

E-mail agents will filter, prioritize, and even reply to routine messages on your behalf. You have to deal with only the few messages that really require your attention.

You Make Better Decisions

Need to choose a health plan, pick a school, negotiate your credit card debt, or refinance your home? Agentic AI can do side-by-side analysis; summarize trade-offs; compare vendors, schools, and banks; and offer recommendations tailored to your values.

You Have a Personal Finance Manager

AI agents will monitor spending, find cheaper subscriptions, shift money between savings accounts and money markets for better interest rates, make investment decisions within your rules, and flag unusual charges on any of your accounts. All the work that AI agents do in this regard aims at a single goal: to build your wealth automatically. Perhaps your motto will be "sleep and grow rich." One can hope!

You Get Fully Personalized Learning

Forget cookie-cutter online courses. Your AI learning coach will guide you through content at your pace and in your preferred learning style, quiz you, recommend resources, and adjust while your knowledge grows. It can also prepare you for a variety of tests from the Law School Admission Test (LSAT) to professional certifications.

You Don't Have to Shop

Tell your agent what you want, such as, "Find me a summer outfit under $100 that fits my style," and it will do all the browsing, comparing, buying, and shipping.

You might also say something like,

> "Rent designer wear for the evening company gala and buy other appropriate clothing for my entire business trip. Have it all delivered to my hotel in time for my arrival. Buy as few outfits as possible that I can mix in different ways to cover my entire stay. Include directions on how many ways I can wear them. Arrange for the rental attire to be picked up and the rest sent to my home when my trip is done."

You may never visit a retail site or pack a suitcase again.

You Get a Travel Planner, Tour Guide, and Concierge

Planning a trip? Your AI will handle flights, hotels, rental cars or other ground transportation, dining reservations, entertainment tickets, and even real-time language translation while you're abroad — and all of it customized to your preferences and budget.

Your Digital Presence Manages Itself

Your online profiles, job applications, and content (such as portfolios or resumes) will stay current without you having to do a thing. Your AI agent will update them while your career evolves and even apply to jobs for you that it knows you'd love.

IN THIS CHAPTER

» Understanding nuance or subtle context

» Handling the truly unknown (or unpredictable)

» Recognizing the full scope of a problem before leaping in

» Making fast, high-stakes decisions under pressure

» Showing genuine empathy or humanity

Chapter **15**

Ten Things Agentic AI Is Terrible at Doing

A gentic AI promises to revolutionize daily life by taking over routine tasks and decision-making. But it's far from flawless. These systems may be powerful, but they still struggle with ambiguity, nuance, and human complexity.

If you know where Agentic AI is likely to fall short, you can set realistic expectations and remember why human oversight still matters. This chapter talks about ten things that Agentic AI can't do well (or at all) — at least, for the foreseeable future.

Understanding Human Emotion in Context

Agentic AI may misread tone, sarcasm, or subtle cues in conversation, leading to awkward or insensitive actions, especially in emotionally charged or personal situations.

Making Moral or Ethical Judgments

Even the most advanced AI agents struggle with gray areas. When values or ethics are in conflict, AI lacks the human capacity for empathy, cultural awareness, or moral reasoning.

Handling Novel, Unstructured Problems

When facing something truly new or outside its training, Agentic AI can flail or even make up stuff. An AI agent can offer irrelevant or offensive answers, *hallucinated facts* (false or misleading information presented as fact), or take illogical actions.

Using Creative Intuition and Artistic Vision

AI can remix existing styles and generate content, but it lacks the lived experience, emotional depth, and abstract intuition that human creators bring to art, music, or storytelling.

If you use AI to create art, remember that you're the creator — the agent is the tool that you use.

Understanding the Bigger Picture

AI agents excel at task execution but may miss long-term implications, strategic context, or systemic consequences, especially in complex domains such as politics or social change. Be diligent in overseeing what an agentic system does in your name because you, not the AI, will suffer any consequences.

Reacting to Sudden, High-Stakes Emergencies

In real-world crises such as a natural disaster, health emergency, or system failure, Agentic AI may respond too slowly or too literally — without compensating for the presence of chaotic human factors or the urgency of life-threatening situations.

Balancing Competing Human Preferences

Agents may optimize for one person's goals while unintentionally leaving out others, such as overbooking shared resources or making decisions that don't account for group dynamics.

If you'd rather share a month-long cruise ship cabin with your spouse than your mother-in-law, make that preference clear to the AI agent and follow up to make sure it followed through.

Building Human Trust and Rapport

Even if it uses a friendly tone, AI can feel robotic or transactional. Only humans can build genuine trust, credibility, and emotional connection naturally.

Respecting Privacy and Boundaries

Without careful programming and guardrails, agents may over-collect data, make invasive suggestions, or take actions that users find uncomfortable or creepy.

Saying "I Don't Know"

AI agentic systems tend to fill in gaps with guesses that can be misleading or flat-out wrong. Agentic AI at the time of writing struggles to admit uncertainty and can elect to act as though it does know what to do.

IN THIS CHAPTER

» Assisting with normal daily life

» Changing online interactions

» Redefining who does what work

» Driving new legal and ethical guidelines

» Reshaping education and healthcare

Chapter **16**

Ten Bold Predictions about the Future of Agentic AI

A gentic AI is more than a tech trend, they are forging a massive transformation in how people live, work, and interact with the digital world. While these autonomous systems evolve from helpful assistants to autonomous decision-making agents acting in our behalf, industries, institutions, and everyday life will all feel this transition's ripple effects. Looking ahead, this chapter offers my ten bold predictions for how Agentic AI will reshape the next decade.

Personal AI Agents Will Become as Common as Smartphones

Within 10 years, and I suspect sooner rather than later, most people in developed countries will have a persistent, personalized AI agent that handles everything from finances, to shopping, to communication. They'll become the default interface that people use to access the internet and business or personal data storage, and they may even take up residence right on your smartphone where they'll push apps to the background.

Agentic AI Ecosystems Will Replace Traditional Apps

Rather than hopping between apps, users will delegate tasks to agents that navigate services behind the scenes. App stores may give way to agent marketplaces, where developers offer agents with new capabilities, not apps. And you will likely soon see apps disappear from things like smartphones and smart TVs, replaced by only an AI interface.

A-Commerce Will Surpass E-Commerce

Agentic AI shopping, where AI agents find, negotiate, and purchase items for you, will outpace manual online shopping in volume, especially for routine purchases and services. This is vastly different from making a standing order on a website for repeat orders, like people do now. Instead, the agents will reorder for you, or seek new purchases based on your calendar such as a gift for an upcoming birthday, a dress for an upcoming party, or wide-legged pants for your upcoming foot surgery.

AI agents will shape themselves to look exactly like you so that you can try on fashions virtually before you buy them. They will act as your personal shopper so that you don't have to shop on various websites by using search engines to find what you're looking for.

SEO Will Evolve into AI Agent Optimization

Instead of optimizing online content for human clicks, businesses will optimize for AI comprehension and compatibility. Agent-readable data, accessed through a standardized protocol like Model Context Protocol (MCP) or *application programming interfaces* (APIs), which govern how agents interact with the data and systems, and *trust signals* (data presented by brands or reliable third parties to prove their worth) will define success in the digital marketplace.

Agentic AI Will Reshape Knowledge Work

From law, to research, to finance, autonomous AI agents will handle large portions of white-collar workflows. Human workers will focus on *oversight* of the agentic systems that perform the work, *strategy* to provide big-picture guidance toward goals, and *ethics* to ensure the veracity of outcomes and the safe and appropriate use of data. And human workers will spend a lot less time manipulating spreadsheets and following e-mail chains.

New Legal and Ethical Frameworks Will Emerge for AI Agents

Governments will create regulations that define what AI agents can legally do, how their decisions are to be audited for compliance, and whom to hold responsible when agents fail, similar to early debates around corporate liability for other types of products.

Insurance companies may even offer AI agent insurance — in case your agent goes rogue or gets attacked. Expect premiums to be high at least until Agentic AI fully matures.

Agentic Education Will Become Mainstream

Students of all ages will have AI tutors that adapt to each student's pace, learning style, and goals. These new personalized education experiences will vastly outperform today's classroom averages.

Students can complete their education at their own pace, even if that's on a slower or accelerated time schedules, and collect skill certificates on their level of competency when they complete courses. Traditional events, such as proms and graduations, may fall to the wayside because student age and experience level are likely to vary greatly, so students in particular course tracks may have little in common to allow for the traditional rites of passage.

Digital Labor Markets for Agents Will Form

People will hire agents to perform specific roles, such as virtual receptionist, data analyst, marketing coordinator, and more. The employer will rate or review the agent's performance, creating a marketplace of AI labor. Yes — just like when you hire a human employee — you can send your AI agent to work and earn an income or a business profit while you go do something fun.

Healthcare Will See a Surge in AI-Driven Preventive Care

Personal health agents will monitor vital bodily functions (like blood pressure and temperature), habits (such as exercising, or not), and risks from your known health conditions. These agents can then schedule tests and flag early warning signs, all while reducing healthcare costs and improving patient outcomes by catching issues before they escalate.

Humans Using Special AI Agents Will Lead Major Social Movements

Activists, creators, and communities will use AI agents to organize, mobilize, and scale causes at speeds and levels of coordination never seen before in an unprecedented level of blending human purpose with AI capability.

For example, I can foresee unemployed people forming ad hoc teams to provide business services that can possibly disrupt their former employers. Further ad hoc teams can form on the fly to make fast money from things like AI agent training, repairs, augmentation, and takeovers for attack purposes. They can disband as quickly and easily as they form. Further, people may use AI in capitalist societies to create disruptive competitors for industries they feel fail society or their demographic, such as health insurance in the U.S.

Appendix

In this appendix, you find out how to assess your business for Agentic AI readiness and get an overview of the phases involved in developing, deploying, and overseeing Agentic AI systems.

Agentic AI Readiness Checklist

To determine whether your business is ready to devise and implement Agentic AI systems, you must examine several areas of the existing business environment. Check out Chapter 5 that discusses how to plan for shifting business functions to Agentic AI systems, and ask yourself the questions contained in this table.

Business Environment Feature	Ask These Questions to Assess Agentic AI Readiness
Process maturity	Are core business processes documented end-to-end, including exceptions?
	Has the business identified agentic opportunities in workflows where autonomy adds value?
	Are dependencies between departments and systems mapped?
	Does the business have a consistent standard for updating and improving process documentation?
Technology and systems	Do core systems offer application programming interface (API) or event-driven integrations?
	Is your infrastructure scalable enough for real-time decision and action loops?
	Are system latencies and bandwidth sufficient for continuous agent operation?
	Are modular architectures in place to allow agents to plug into existing workflows without breaking them?
Data readiness	Is critical operational data accurate, clean, and regularly refreshed?
	Can agents access the data in real time without manual intervention?
	Is the data semantically consistent across different sources and formats?
	Are security and privacy controls applied to all datasets the agent might access?

Business Environment Feature	Ask These Questions to Assess Agentic AI Readiness
Governance and trust	Has the business clearly defined and approved autonomy boundaries for AI agents?
	Has the business defined escalation protocols that take effect when agent decisions exceed authority limits?
	Does every agent decision and action have an audit trail?
	Do governance processes comply with relevant regulations (for example, the federal Health Insurance Portability and Accountability Act, HIPAA, in healthcare-related systems)?
People and culture	Has the business defined roles and responsibilities for human–agent interactions?
	Has the business trained staff on how to work with and supervise agents?
	Does the business have a feedback process for employees to flag issues or propose AI system improvements?
	Are change-management strategies in place to handle cultural resistance?

Phases for Creating and Using Agentic AI Systems

Follow these phases to define, build, operate, and oversee a regulatory-compliant Agentic AI system for a pilot project. Refer to Chapter 5 for more information about implementing Agentic AI and to Chapter 6 for a look at various use cases.

Phase 1: Strategic foundation and use case selection

Begin by establishing clear, measurable objectives that align with your broader business strategy. If you want a little guidance on where to start, consider that successful pilot objectives typically fall into three categories: operational efficiency gains, cost reduction targets, and capability enhancement metrics. For example, rather than vague goals like "improve customer service," specify "reduce average resolution time by 40 percent while maintaining 95 percent customer satisfaction scores."

The pilot project should demonstrate both the AI agent's autonomous decision-making capabilities and its integration with existing business processes. This dual focus ensures that technical success translates into business value while revealing integration challenges early in the deployment cycle.

Avoid choosing a pilot project that involves high regulatory scrutiny, significant safety implications, or complex stakeholder undercurrents for initial pilot projects. These factors introduce variables that can obscure the AI system's core performance and complicate success evaluation.

Phase 2: Technical architecture and integration planning

Design the pilot architecture with modularity and observability as core elements. Implement comprehensive logging systems that capture not only the AI agent's decisions but also its reasoning process, confidence levels, and the data sources informing each action. This transparency becomes crucial for debugging, compliance, and building organizational trust in autonomous systems. The architecture should support both autonomous AI operation and human intervention capabilities.

Create a vigorous rollback mechanism that can quickly revert to manual processes if the Agentic AI encounters unexpected scenarios or produces unsatisfactory outcomes. This capability reduces organizational anxiety about autonomous deployment and provides confidence for stakeholders during the pilot phase.

Phase 3: Governance and monitoring framework

Establish both quantitative and qualitative metrics that provide a reliable measure of the AI agent's effectiveness across multiple dimensions. Quantitative measures may include task completion rates, accuracy scores, processing speeds, and cost per transaction. Qualitative assessments may evaluate decision quality, stakeholder satisfaction, and alignment with business objectives.

You can establish real-time monitoring dashboards and regular review cycles in which human experts evaluate a sample of the AI agent's decisions and provide feedback that can inform how systems are adjusted and refined. This ongoing evaluation process ensures continuous improvement and maintains quality standards throughout the pilot period.

Phase 4: Risk management and compliance

Conduct comprehensive risk analysis that covers technical failures, business process disruption, data privacy concerns, and regulatory compliance requirements. Consider both immediate operational risks and longer-term strategic implications of autonomous AI deployment, which may include evaluating potential impacts on workforce dynamics, customer relationships, and competitive positioning.

For each identified risk, develop specific mitigation strategies and monitoring protocols that can detect early warning signs before issues escalate. For example, consider these areas:

» **Data governance and privacy:** Implement strict data governance protocols that ensure the Agentic AI operates within established privacy and security boundaries. These protocols include access controls, data retention policies, audit trails, and compliance verification mechanisms that satisfy regulatory requirements while also enabling effective AI operation.

» **Technology complexity when scaling AI:** Building an Agentic AI system is an understandable challenge in software development (see Chapter 10) but deploying systems is a massive infrastructure endeavor that encounters obstacles due in part to integration with other AI systems and various data sources. And through all the integrations, AI agents must work in real-world traffic while also meeting service level agreements (SLAs), budget, and regulatory requirements.

Index

P

pay-as-you-go access, 259

people, agents replacing, 255
  AI takeover perspectives, 258
  business perspective, 255–256
  cost considerations, 256–257
  media narratives, 257

performance optimizers, 156

personal AI agents as common devices, 311–312

personal AI finance management, 305

personal development planning, 282

personalized learning via AI, 305

personalization, 36

personalized treatment agents, 142

personalizing AI learning agents, 160–161

personalizing workflows, 98
  informed actions, 98
  role adaptation, 99
  tailoring output for user or co-agents, 99–100

phases of Agentic AI system development, 316
  governance and monitoring framework, 317
  risk management and compliance, 318
  strategic foundation and use case selection, 316–317
  technical architecture and integration planning, 317

philosophical and existential questions about Agentic AI, 290
  determinism examined, 291–292
  free-will exploration, 290–291
  predictability of AI agents, 292–294

pilot framework design, 126–127

pilot-level use cases, examples of, 125–126

Pinecone, 35

planning, Agentic AI, 116
  five pillars of planning, 117
  reviewing the plan, 119–120
  SMART goals and detailed follow-up, 117–119

planning and goal-conditioned learning, 60

planning and implementing Agentic AI, 120

building or integrating Agentic AI systems, 127–133

business models and value creation, 139–140

expansion and scaling, 135–136

governance and trust, 136–137

high-impact use cases, 124–126

pilot framework design, 126–127

readiness evaluation, 122–124

running, measuring, and refining, 133–135

strategic intent, 121–122

workforce upskilling, 138–139

policy updates, 62

portfolio management, investment research and, 215–216

post-production agents, editing with, 158–159

predictability of AI agents, 292
  clear operations, 292–293
  oversight focus, 293
  proof and security, 294

pre-training models for Agentic AI systems, 237–238

preventive healthcare, AI-driven, 314

privacy and boundary respect, 309

problem-solving, creative, 268

procurement and negotiations with AI agents, 148–150

productivity gains, 209
  current progress and, 210–211
  expected, 211–212

Productivity J-Curve, 212, 299

professional and industry leaders, 231

professional development, 161

project managers, 266

prompt-based systems, traditional, 64

prompt engineering, context engineering vs., 86
  differences in engineering methods, 88–89
  maintaining agent behavior, 95
  need for both practices, 87–88
  prompt engineering in Agentic AI, 93–94
  prompting and tool integration, 94

swarm intelligence and decentralized agency, 296–298

symbolic signals, 59

synthetic agency and collective intelligence, 294

multi-agent ecosystems, collective intelligence in, 295–296

swarm intelligence and decentralized agency, 296–298

system design differences, 107

systems thinking, 267

T

task allocation mechanisms, 60

task decomposition, 30

computational complexity and, 50–51

task delegation strategies, 215

common pitfalls, 217–218

framework for task assignments, 216–217

teaming rather than competing, 215–216

technical architecture and integration planning, 317

technical architecture approaches, 75

agentic frameworks, 76–77

AI agent-building platforms, 77

building from scratch, 76

supporting system development, 77–78

Technical Stuff icon, 4

technologists, roles for, 271–272

technology complexity when scaling AI, 318

theory of mind modeling, 59

three-dimensional (3D) reasoning capabilities, 241

timing issues, 107

Tip icon, 3

tool linkage and use, 31

tool mediation, 186

traditional apps, Agentic AI ecosystems replacing, 312

traditional automation, Agentic AI and, 260

combining, 261–262

when to choose Agentic AI, 261

when traditional automation works, 260–261

traditional prompt-based systems, 64

training AI agents, 237

built-in protections and ethical practices, 239

pre-training models for Agentic AI systems, 237–238

proactive piece, adding, 238

training example, 240

training in simulated and real environments, 238–239

training in simulated and real environments, 238–239

transition to the Agentic AI workplace, 203

earning opportunities with AI agents, 207

expense reduction via using AI agents, 205–206

financial relief, AI tools for, 205

sophisticated job hunting, using tools for, 204–205

transparency

in Agentic AI reasoning, 179–180

lack of, 107–108

transportation control lessons, 170

travel planning and concierge services, 306

TREEMENT, 144

TrialMatchAI, 144

Trust, Risk, and Security Management (TRiSM), 294

Twisted Sequential Monte Carlo (TSMC), 15

U

uncertainty, admitting, 309

universe-sized datasets, Agentic AI requiring, 262–264

unrealistic mandates, avoiding, 250

upskilling for agentic age, 265

actionable learning strategies, 277–278

AI agent manager, becoming, 266

creative problem-solving, 268

creative professionals, roles for, 272–274

data literacy, 267

emotional intelligence vs intuitive intelligence, 283

ethical fluency, understanding, 280–281

ethics and governance, 268
future-proofing careers, 279–280
human judgment, value of, 281–282
leaders, roles for, 274–275
lifelong learning and agile adaptation, 275–280
personal development planning, 282
reframing skills and habits, 269–271
strategic learning framework, 276–277
systems thinking, 267
technologists, roles for, 271–272
user experience (UX), 158
U.S. National Institute of Standards and Technology, 181

V

value alignment, collective effort toward, 178–179
value creation, business models and, 139–140
value learning drift, 174–175
vector embeddings, 35
VectorStoreRetrieverMemory, 35
vehicle-to-everything (V2X) interactions, 54
video editing, 159

visual exploration, 158
voice, intent, and semantic interfaces, 95
    expressing intent with AI, 95–96
    semantic interpretation, 96–97

W

Warning icon, 4
web, Agentic AI, 21
    expanding online duties, 22
    scaling from citywide to global, 22–23
website content, structure and focus of, 26
workforce upskilling, 138–139
World Economic Forum (WEF), 235
world modeling, 56–57
world models beyond data training, 240
    existing world-model systems, 242–243
    operation in 3D environments, 241–242
    pros and cons, 243–244
    real-world learning, 242
writing agents, 157

About the Author

Pam Baker is an award-winning journalist, author, and instructor specializing in artificial intelligence (AI) and emerging technologies. She teaches widely attended AI courses on LinkedIn Learning and is the bestselling author of *Generative AI For Dummies* and *ChatGPT For Dummies* (1st and 2nd editions). Her additional works include *Decision Intelligence For Dummies* (Wiley) and *Data Divination: Big Data Strategies* (Cengage), among other titles exploring the business and societal impact of advanced technologies.

Her reporting and analysis have appeared in leading outlets such as Institutional Investor, Ars Technica, CIO, CISO, InformationWeek, CNN, PC Magazine, TechTarget, and Dark Reading. Baker covers breaking news as a stringer for *The New York Times.* She's also a sought-after speaker at science and technology conferences; her presentation on mobile health data and analytics was published in the *Annals of the New York Academy of Sciences.*

Earlier in her career, Baker conducted research and analysis for firms including ABI Research, VisionGain, and Evans Research. She's a member of the National Press Club (NPC), Women's Media Group (WMG), and the Internet Press Guild (IPG). You can find out more about Baker's work and credentials, as well writing samples, on her LinkedIn page (www.linkedin.com/in/pambaker).

Dedication

For Stephanie Baker Forston and David Forston, Ben Baker and Dr. Katherine Poruk Baker, and my endlessly delightful granddaughter crew — Mirabel, Coco, Poppy, and Charlotte — thank you for filling my life with inspiration, joy, and sustaining love.

Ben, your thoughtful feedback and technical guidance once again kept me steady as I navigated the fast-moving world of Agentic AI. Your support means everything. Stephanie, your rare blend of strategic insight and creative skill never cease to inspire me. Your support and energy mean more to me than you know. Katherine, your seaside office has become something of a lucky charm — four books strong now — with Luna and Cinny loyally standing guard as only cats can. Your support is both practical and inspiring!

To all of you: I could not have carried this work across the finish line without your support, encouragement, and the happiness you bring to my world. Thank you all for so patiently supporting me through yet another writing marathon.

Author's Acknowledgments

Creating a book is always a monumental effort, demanding the talent and dedication of many skilled and creative people. This project, however, tested us all in a variety of ways: the technology was (and is) still emerging, evolving at breakneck speed, and the deadlines were relentless. Yet through it all, the team rose to the challenge, ensuring that this book could meet the moment and serve readers well. I am profoundly grateful to everyone whose expertise, hard work, and support shaped these pages and elevated them far beyond what I could have achieved on my own.

Special thanks to Steven Hayes for making this book possible. Many thanks to Leah Michael and Laura K. Miller for their steady guiding hands throughout this marathon. Hugs and kudos to Leslie Lee for her remarkable AI knowledge and her sharp eye in the technical editing of this book, as well as for her own accomplishments in Agentic AI in the real world. And of course, and always, a heartfelt thanks to my agent, Carole Jelen.

Publisher's Acknowledgments

Executive Editor: Steve Hayes

Development Editor: Leah Michael

Copy Editor: Laura K. Miller

Technical Editor: Leslie Lee

Managing Editor: Sofia Malik

Production Editor: Magesh Elangovan

Cover Image: © InfiniteFlow/stock.adobe.com